Pauline Reilly has researched and written about penguins for many years. She spent thirteen years on Phillip Island and along the Victorian coast-line, mounting the first intensive campaign in Australia to band and thus conserve the Little Penguin. She subsequently spent several months studying other penguin species on Macquarie Island. A Past President and Fellow of the Royal Australasian Ornithologists Union, she has written many books, including *Fairy Penguins and Earthy People*, and a children's book on penguins. Pauline Reilly lives on the Great Ocean Road in south-western Victoria with her husband of fifty years and is the mother of three sons.

## HANDBOOK OF AUSTRALIAN, NEW ZEALAND AND ANTARCTIC BIRDS

The *Handbook of Australian, New Zealand and Antarctic Birds* (HANZAB) is one of the most significant and exciting projects in international ornithology. HANZAB provides an up-to-date, accurate and comprehensive synthesis of our knowledge of all the birds that occur in this region. This six-volume project continues the tradition of authoritative publications on Australasian birds that began with Gould's *Handbook of the Birds of Australia* (1865). It will have an enormous impact on the future direction of research and on the conservation of Australasian and Antarctic birds.

Already available
Volume One — *Ratites to Ducks*
Volume Two — *Raptors to Lapwings*

To be published 1995
Volume Three — *Pratincoles to Pigeons*

# PENGUINS
## OF THE WORLD

# PENGUINS
## OF THE WORLD

Pauline Reilly

Oxford University Press

OXFORD UNIVERSITY PRESS AUSTRALIA

Oxford New York Toronto
Delhi Bombay Calcutta Madras Karachi
Kuala Lumpur Singapore Hong Kong Tokyo
Nairobi Dar es Salaam Cape Town
Melbourne Auckland Madrid
and associated companies in
Berlin Ibadan

OXFORD is a trade mark of Oxford University Press

National Library of Australia
Cataloguing-in-Publication data:

Reilly, Pauline, 1918–
    Penguins of the world.

    Bibliography.
    ISBN 0 19 553547 2.

    1. Penguins. I. Title.

598.441

Edited by Stephen Dobney
Designed by Derrick I Stone
Cover photograph from Jonathan Chester
Typeset by Derrick I Stone
Printed and bound by Impact Printing Victoria Pty Ltd.

Published by Oxford University Press,
253 Normanby Road, South Melbourne, Australia

# Contents

# Preface

To see penguins coming out of the sea — walking, scrambling, leaping — is to become enchanted. For some people that enchantment may need reinforcement, but right from the start I was bewitched irrevocably. The Little Penguins on Phillip Island in Victoria began my captivation and the penguins on Macquarie Island in the subantarctic strengthened the bonds.

I can clearly recall seeing for the first time the amazing spectacle of hundreds of thousands of penguins in one panoramic view on Macquarie Island. When I sat on the beach among a group of resting Royal Penguins, I then 'became a penguin', the birds rearranging themselves a pecking distance away from me. They were too close for my camera. As I leaned back I started a wave of movement, those in front leaning towards me, those behind away. They showed no apparent distress at my presence, but I now know that it would have been better to have kept well away.

That one week spent with penguins on remote and beautiful Macquarie Island made me determined to return. I did so the following summer and spent three glorious months grappling with Gentoo Penguins. Their lack of enthusiasm for me was expressed by yelling, scratching, biting, thrashing me with their flippers and constantly splattering me with guano. After crawling up a steep slope through tussock grass to examine the contents of their nests, I found it expedient to wait with face averted until the attendant bird had turned its back and let loose a squirt of sticky, smelly guano, an intimidating method of defence.

I have met most of the world's species of penguins, some of

them only in captivity. Though there are similarities between them, belonging as they do to the same family, there is also a wonderful diversity in the manner in which they have adapted to their environments, from the equator to the Antarctic.

Not so long ago, when it was my privilege to open the Second International Penguin Conference, I urged those giving papers to remember that they were speaking to many who were not knowledgeable in the more esoteric aspects of penguin biology. That thought has been very much in the forefront of my mind during the writing of this book. *Penguins of the World* has been written to give the non-specialist reader and naturalist an insight into penguin biology without having to wade through endless pages of unwanted information. I have tried to convey accurately the world of penguins in simple language, the language most of us

Fig. 1. *Left* Royal Penguins nesting at Macquarie Island.
Fig. 2. *Above* The author on Macquarie Island with King Penguins and baby elephant seal.

enjoy when we read for pleasure. The more erudite reader may find some explanations redundant, such as those dealing with genera and species, though such explanation is kept to a minimum.

I gratefully acknowledge that the source of the illustrations, figures and maps, and much of the information for *Penguins of the World* is the *Handbook of Australian, New Zealand and Antarctic Birds* (HANZAB) published by Oxford University Press in 1990. The bibliography includes the other main sources and papers published more recently.

I have many people to thank for their assistance and encouragement. My thanks go to all those who, bewitched by penguins, have carried out research under extreme climatic conditions that entailed severe physical hardship.

I thank my referees, Dr P. Dee Boersma, Dr R. J. M. Crawford, Dr J. P. Croxall, Prof. J. M. Cullen, P. Dann, J. T. Darby, B. Dyer, Dr K. Kerry, Dr Y. Le Maho, G. Moore, Dr P. J. Moors, G. Robertson and E. J. Woehler, all of whom commented with great thoroughness on drafts of the manuscript. Their up-to-date information replaced much that was originally in *HANZAB*. I hope they will not be too disappointed at my inability to incorporate all their suggestions. Had I done so, the length would have been excessive and the content unbalanced. Any errors that remain must be attributed to me alone. I thank also Dr J. Waas for additional information, Peter Higgins (co-editor of *HANZAB*) for casting an RAOU eye over the manuscript, and Dr A. B. Pittock, who clarified my foggy ideas about the ozone layer and global warming.

PR

## Chapter 1

# The Diversity of Penguins

INTRODUCTION

This general introduction is designed to give a broad picture of penguins. Every statement could be qualified by adding words such as 'most penguins', 'may sometimes', and so on, which would make tiresome reading. Consequently exceptions are stated only where individual species differ markedly from the general picture.

To avoid tedious repetition for each species, much of the penguin's life is covered in this first chapter. If the information you are looking for is not contained in the chapter on that particular species, it is probably here. For example, most but not all penguins lay two eggs, with both parents involved in the whole cycle of breeding from nest building to the fledging of the chick. Where the regime is different, it is described both in the 'Introduction' to the genus, and for the individual species under 'Breeding'. Words and terms that may be unfamiliar to the general reader are explained in the glossary.

For each genus involving more than one species there is a general introduction at the beginning of the chapter. This is followed by species accounts in which the sections are arranged in the same order: Description; Distribution, Dispersal and Population; At Sea and on Land; Behaviour; Breeding; and Threats and Conservation.

WHAT IS A PENGUIN?

The first seafarers to sail south and discover penguins wondered whether they were beast, bird or fish. They finally settled on 'feathered fish'. Even today, when people first see penguins emerging from the sea, they ponder the same question. The answer is neither beast nor fish, but bird. Penguins are birds because they have feathers and only birds have feathers. Because their legs are set so low down on their bodies, penguins stand upright, and perhaps it is this similarity to ourselves that makes penguins so attractive to people. Unfortunately, the dignified penguin is too often portrayed as the hackneyed caricature of a waiter in evening dress.

## DESCRIPTION

The feathers of penguins are black, blue or grey on the upperparts and white underneath. Some penguins have crests and various colours on the head and bill. Flippers are dark above and the white on the underside is marked with various spots and blotches. The overlapping feathers are stiff and tightly packed above a layer of fine down. When penguins enter the water the feathers press down tightly to keep water out. Some of the air trapped under the feathers is gradually expelled and rises in a stream of bubbles as the bird swims.

Penguins' bodies have thick deposits of fat which increase greatly as a prelude to moult. They are well insulated against the cold by their fat, their down and the air trapped under their feathers. In fact, when it is at all warm, even in the Antarctic, they have trouble losing heat. Like other birds they cannot sweat as mammals do, but have to lose heat by fluffing up their feathers, by panting or by shunting blood to the less insulated parts of their bodies: inside their mouths, underneath their flippers, and their legs and feet.

It is generally stated, though apparently without proof, that the penguin's colour helps to camouflage it when it lies on the surface of the water. When an underwater predator looks upward, the penguin's light-coloured underparts blend better into the lighter colour of the sky above. Conversely, when an aerial predator looks down, the dark back blends into the colour of the sea. When the penguin itself becomes a predator and searches for prey, this camouflage acts in its favour.

Penguin chicks are covered in short down when hatched (protoptile plumage). This is soon replaced by a thicker coat of down (mesoptile plumage). This second coat varies greatly in colour between the genera and within each genus. For instance in the *Aptenodytes* genus where the adult plumage of both species is similar, Emperor chicks are grey with a black mask, while King chicks are brown all over. In the *Pygoscelis* genus, Adelie chicks are dark grey, Chinstrap chicks have grey backs and faces and are pale grey beneath, and Gentoo chicks have black backs and faces and white underparts. The differences between the chicks in the other genera are less marked and they are generally grey-brown in colour.

The young at fledging are a little smaller than the adults and, after their second down has moulted, their plumage mostly resembles that of the adult.

Fig. 3. Little Penguin chick in second down.

## PREENING

Penguin chicks preen as soon as their feathers begin to grow, and this becomes a constant imperative throughout their lives. It is essential that feathers be maintained in the best possible condition to keep them waterproof and to keep out the cold. When a penguin comes ashore, it preens; when it is standing around with nothing much else to do, or is irritated by parasites such as ticks and fleas, again it preens. In doing so it squeezes a waxy substance from the uropygial (preen) gland at the base of its tail and spreads this substance through all its feathers by running them through and under its bill. To preen around the head and eyes the bird has to scratch with its feet, but another penguin does the job much more effectively. Preening of one bird by another is termed allopreening. Not all penguin species allopreen.

Fig. 4. Little Penguin preening. (Note the preen gland.)

Fig. 5. Royal Penguins allopreening.

Allopreening is usually regarded as both comfort behaviour, to remove parasites, and social behaviour, to maintain the pair bond. It is common in the penguins of temperate and equatorial waters where parasites abound, but it is not recorded for some of the penguins that breed in the colder regions. Emperor Penguins do not allopreen, nor do they appear to have external parasites.

Ticks, for instance, would have no chance of survival on the ice (they need to drop off their host into a sheltered environment to maintain their cycle). The pygoscelids (Gentoos, Adelies, Chinstraps) do not allopreen, but the King Penguin does. The distribution of King Penguins overlaps that of Gentoo Penguins that do not allopreen and the subantarctic eudyptids (crested penguins) that do. There is a lack of information on parasites for some of these species, so it cannot be concluded that allopreening is activated by the presence of parasites. It may be more a form of social behaviour rather than plumage maintenance.

## FOSSIL PENGUINS

Until recently, the oldest fossil penguins were estimated to be 45–50 million years old, but a recent find in New Zealand may be 55 million years old. The ancestors of penguins were flying birds. Although penguins cannot fly through the air, they do 'fly' under water. Their wings have been highly modified into paddles, referred to as flippers, and their long bones are not light and filled with air like those of flying birds. Penguins are shaped like torpedoes and spend most of their time in the sea, but they still have to come ashore to breed and to moult.

## DISTRIBUTION

Penguins live in the southern half of the world, from the icy Antarctic continent in the south — the coldest, windiest and driest place in the world — to the Galapagos Islands where the Galapagos Penguin breeds on both sides of the equator. The Antarctic is that region south of the Antarctic Polar Front. Along this front there is an upwelling of nutrient-rich cold water as it meets the more temperate subantarctic waters.

In the minds of many people, penguins are associated with the Antarctic, all migrating there at some time. In fact, more than half the number of penguin species never go to the Antarctic. Another misconception is that only two species, Emperor and Adelie Penguins, breed on the Antarctic continent. In reality,

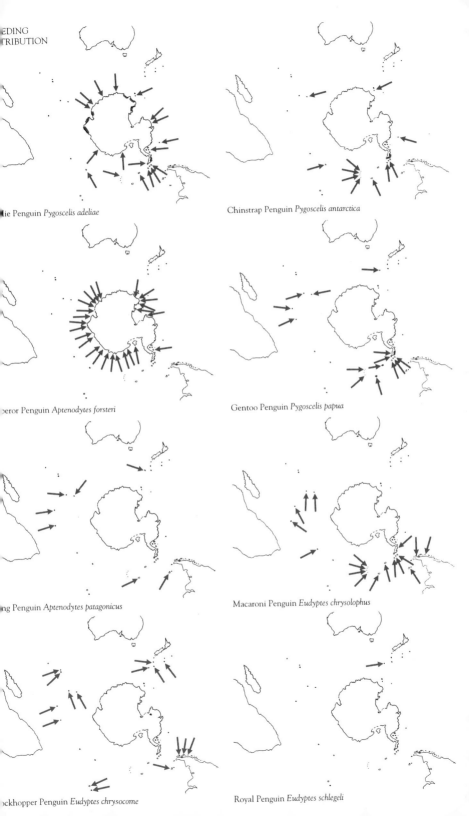

lie Penguin *Pygoscelis adeliae*

Chinstrap Penguin *Pygoscelis antarctica*

peror Penguin *Aptenodytes forsteri*

Gentoo Penguin *Pygoscelis papua*

ng Penguin *Aptenodytes patagonicus*

Macaroni Penguin *Eudyptes chrysolophus*

ckhopper Penguin *Eudyptes chrysocome*

Royal Penguin *Eudyptes schlegeli*

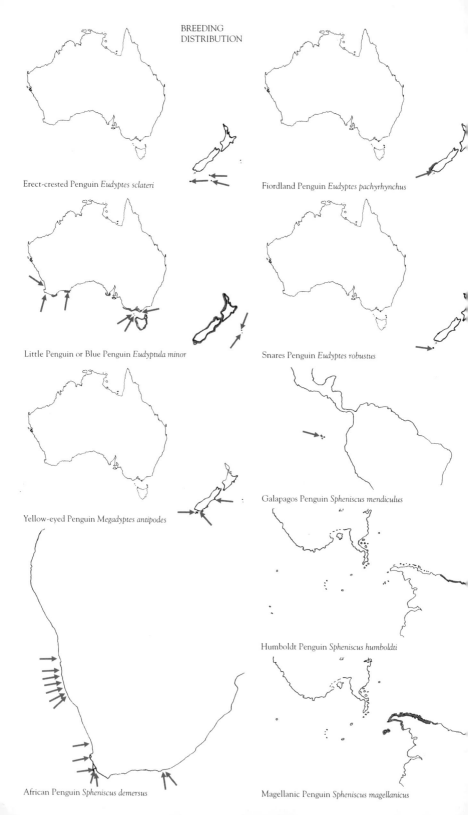

BREEDING
DISTRIBUTION

Erect-crested Penguin *Eudyptes sclateri*

Fiordland Penguin *Eudyptes pachyrhynchus*

Little Penguin or Blue Penguin *Eudyptula minor*

Snares Penguin *Eudyptes robustus*

Yellow-eyed Penguin *Megadyptes antipodes*

Galapagos Penguin *Spheniscus mendiculus*

Humboldt Penguin *Spheniscus humboldti*

African Penguin *Spheniscus demersus*

Magellanic Penguin *Spheniscus magellanicus*

Gentoo and Chinstrap Penguins are interspersed with Adelie Penguins on the Antarctic Peninsula, making a total of four species breeding on the Antarctic continent.

## RELATIONS

All 17 species of penguin (or 18, see Rockhoppers, page 74) are grouped within a single family, the Spheniscidae. The family name is derived from *sphen*, a Greek word meaning 'wedge', which refers to the shape of the penguin flipper. Within this family there are six groups or genera (a single group is a genus). Within each genus there may be one or more species. Each bird has a scientific name composed of two parts: a generic and a specific. The generic name indicates that those within the group are closely related and the specific name separates one species from another. Sub-species or races are not dealt with in this book. The closest relatives of penguins today are the petrels and shearwaters, all of which can fly.

The name 'penguin' was first given to a completely different seabird family, the auks of the northern hemisphere. Probably the best known and the most photographed of these is the puffin. Auks are not related to penguins, although they look like them. Auks are dark brown and white birds that stand upright but, unlike penguins, they can fly through the air as well as through the water. An exception was the largest of the auks, the great auk, which could not fly. It was easy to catch and both the bird and its eggs were eaten. As a result, the great auk became extinct late last century.

Some species of penguin have been studied in great detail, but others, such as the Humboldt Penguin that lives on the coasts of Peru and Chile, have not. Consequently, there is less information available for some species.

## POPULATION SIZE

It is extremely difficult to determine the whole population of any penguin species anywhere. The whole population is not ashore at the one time and pre-breeders are not accounted for when breeding populations are surveyed. Should chicks be included in such surveys, even though many will not survive to maturity? Those penguins that breed over a long period and do not all lay at the same time, or those that breed in burrows, in small caves, in thick vegetation or in crevices are impossible to count accurately. Some will be ashore, while others may be at sea for days.

Breeding Emperor Penguins can be counted reasonably accurately by using aerial photography after the females have laid their eggs and departed, leaving only incubating males. Another method is to count the number of individuals standing around on the ice during moult. With any counting method, even with species that breed synchronously, there is no certainty that one member of every breeding pair is ashore at any one time.

Despite all these difficulties, an attempt was made in 1993 to estimate the population of penguin pairs breeding south of the Antarctic Polar Front by examining the most up-to-date population figures. Bracketed figures indicate the percentage that breed in the higher (colder) latitudes south of the front, the remainder breeding north of it in the lower (warmer) latitudes: Emperor 195000 (100%); King 1.02 million (95%); Adelie 2.47 million (100%); Chinstrap 7.49 million (100%); Gentoo 236000 (75%); Macaroni 11.72 million (99%); Rockhopper 738000 (20%); giving a total of 23.87 million pairs of penguins. These are *minimum* estimates. If pre-breeders, non-breeders and failed breeders are included, the estimate might need to be increased by 40 per cent. This count, of course, excludes all those penguins (ten species) that are never in the Antarctic.

## MEASUREMENTS

Measurements are essential at all stages of bird study to learn about a bird's development and to assess the effects of seasonal and environmental changes year by year. But measuring penguins is not easy. They do not co-operate by lying flat and still against a measuring scale. Instead they scratch, bite, beat with their flippers and excrete all over the field worker. They are restrained for weighing in a bag or similar device. Measurement of length is from the tip of the bill to the tip of the tail. This is more accurate than the height of a standing bird because its height varies depending on what it is doing. This means that to get the height of a standing bird a few centimetres needs to be deducted from the lengths shown in bird books which often represent the bill to tail measurement of a museum specimen.

The measurements given in each of the species descriptions should not be regarded as absolute but rather as a guide to the size of the bird. Confusion may arise in the case of birds with long tails, such as the pygoscelids (Gentoos, Adelies and Chinstraps). Compared with a shorter-tailed bird, a standing pygoscelid is shorter than its measurements might suggest. For example, the Yellow-eyed Penguin actually stands taller than the Gentoo Penguin although its bill to tail measurement is smaller. Where the feet protrude further than the tail, measurements are sometimes taken from bill to feet.

Male penguins are generally larger than females in all measurements except weight (mass), which varies for both sexes according to the season. Just before egg-laying, females of some genera are heavier than males, but the males quickly become heavier once they go to sea and replenish their reserves. All birds are much heavier at the start of fasting and much lighter at its conclusion, such as when moulting or when incubating eggs for long periods. For instance, a male Emperor at the beginning of breeding may weigh more than 40 kg, reducing to 23 kg by the end.

## NAVIGATION

An idea of how penguins navigate was gained from experiments with Adelies on the Antarctic continent. During the chick-rearing period, males that were defending nest sites but without chicks were transported to two areas that lacked any landmarks, such as mountains, that could be used for orientation. They were then released at distances of 340 km and 1500 km from their place of collection.

Birds were followed by telescope until out of sight at 3–4 km. Those released under cloudless skies quite quickly oriented themselves and set off in a straight line, alternately tobogganing and running towards the coast and ice-covered sea. Under veiled sun their direction was less definite, and when the sun was obscured the direction taken was erratic.

The majority of birds released at both distances and with sufficient time for return to their original colony, did indeed do so. The conclusion reached was that penguins navigate by the sun and use an inbuilt biological clock to compensate for its movement across the sky (the changing sun azimuth positions).

## LOCOMOTION ON LAND

Penguins walk with a waddling gait which looks slow and cumbersome, but they can also hop and run. When jumping off a rock or over a crevice they lean forward to assess the distance to be jumped. They can jump up to almost their own height, measuring it by placing their bills on the ledge to which they will jump. When passing the territory of other penguins in a tightly packed colony, they sleek themselves down with neck extended, head high and flippers to the sides, trying to take as little space as possible and thus avoid being pecked. On hard surfaces they walk, but on soft snow they toboggan, using their feet to push along and to act as brakes. Uphill they use their flippers to propel themselves. They can toboggan faster than they can walk, even on sand.

## DIVING AND LIFE AT SEA

In the water penguins are swift and graceful, travelling at 7–8 km/h, though capable of more than 12 km/h in short bursts when pursuing a school of fish or being chased by a predator. Under water they use their flippers with much the same action as birds that fly through the air. Their tails and feet (the three front toes are webbed) are used only as rudders for steering (and possibly braking). When moving quickly, either as pursuer or pursued, some species break through the surface of the water in an action

Fig. 6. Adelie Penguins on the move.

referred to as 'porpoising'. Actually the penguin looks more like a flying fish as it erupts from the water and travels for a short distance above the surface. This porpoising action enables the birds to catch a quick breath before re-entering the water to swim not far below the surface. On the surface they swim slowly with a small flipper stroke. A large stroke would bring the flipper out of the water and therefore would not give any thrust.

All penguins dive. It is easier for larger penguins to dive deeply because they have greater reserves of oxygen relative to their muscle needs, but even the Little Penguin has been recorded diving to nearly 70 m. Most dives are less than 100 m in depth, but King and Emperor Penguins often dive deeper. The deepest dive recorded is more than 400 m for the largest of all penguins, the Emperor. It has also been recorded staying under water for 18

Fig. 7. Adelie Penguin leaping into the water.

minutes, though the duration of most of its dives is seldom more than eight minutes. In deep dives there is no illumination, so prey capture cannot be visual. Only the penguins themselves know how they capture prey at these depths.

Penguins at sea use different methods of travel, depending on what they are doing. When travelling to and from feeding areas they proceed in short, shallow dives of about 20 seconds, followed by a rest on the surface for about the same length of time. When searching for prey they dive steeply and deeply and, having located it, they spend longer under water feeding. Their natural buoyancy helps bring them back to the surface again. Studies to date do not explain how penguins avoid the normal hazards of deep diving (decompression sickness and nitrogen narcosis).

Fig. 8. King Penguins at sea.

Penguins pursue and swallow their prey under water, and feed either as individuals or in groups. There is no evidence of systematic co-operation between individuals, although it is suggested that the spheniscids (the four species with black chest bands) do feed co-operatively. In pursuit dives an individual penguin swims around a school of fish until they are packed tightly together. It then dives into the pack from underneath because it can see better looking upwards than downwards. At other times it chases single fish.

Inside the penguin's mouth are gristly projections, a little like soft, backward-sloping teeth, which enable the birds to hold their live and wriggling prey.

## VISION

The penguin's eyes are equipped for vision under water as well as on land. Penguins, like most birds, have binocular vision (they look forward with both eyes) with some overlap of the field of vision of each eye. This probably helps them to judge the distance of objects in front of them, such as prey which they grasp with their bills. Their eyes are also large and adapted to the poor light conditions they encounter under water.

## FOOD

Penguins feed in cool waters, even on the equator. They feed on three types of food: small fish, squid (cephalopods) and krill (crustaceans). The depth to which a penguin dives determines what prey is available to it. Antarctic krill, for instance, swim deeper in the water during the day and rise towards the surface at night. The larger the krill, the fewer are needed to make a meal, with gravid (egg-carrying) female krill having a greater energy value than males. Some penguins take all three types of prey; others vary their diet depending on their location.

Penguins do not need fresh water to drink, although some eat snow and drink from puddles. Their kidneys and nasal glands excrete excess salt from the sea water taken in with food. When they shake their heads, little salty droplets spray out from their nostrils.

Unless there is some specific food difference, such as the manner of catching or its implication in conservation, food is not dealt with specifically in later chapters.

## FOOD CONSUMPTION

The total food consumption of the estimated 23.87 million pairs of penguins breeding south of the Antarctic Polar Front amounted to 18.1 million tonnes annually, composed of 3.7 million tonnes of fish, 13.9 million tonnes of krill and 539000 tonnes of squid. Again, these are estimates of the *minimum* consumption of marine resources in the seas surrounding Antarctica.

## DISPERSAL AND STUDY AIDS

Some penguins travel long distances when they are not breeding, but it is not easy to find out where they go at sea. Studies using a metal band placed around penguin flippers or (in the past) on legs, have been carried out for many years to find out how far birds travel, whether they stay together as breeding pairs and how long they live. Juveniles in their first few years of life stay away from their breeding grounds, but they must come ashore to moult. An examination of museum specimens of vagrant penguins (those that are a long way out of their usual range) showed their toenails to be long and curled. Toenails abrade in contact with rocks and the ground surface, so this excessive length indicated that the penguins had been at sea for a long time without making landfall.

Recoveries of bands on dead penguins are made by people who walk along beaches in the temperate regions. By removing the bands and sending them to the address inscribed on them, they will learn where and when that bird was banded.

More complex technology, including electronic devices and satellite tracking, is now enabling the life of penguins at sea to be studied in greater detail. One system automatically weighs, identifies and determines the direction (in or out of the colony) of penguins as they walk across a weighing platform. Thus the duration of foraging trips, length of incubation shifts, and the weight of food carried to young can be determined. An implanted electronic identification tag (a passive transponder) is read by an electronic tag reader at the weighbridge. Data are transferred automatically via satellite to the headquarters of the research organisation on another continent.

Another system uses backpacks that contain a depth recorder, compass and speedometer to record the diving depth, direction and length of trip and the penguin's speed at regular intervals. Stomach recorders indicate whether prey is taken in a particular dive by noting the change of temperature inside the stomach when a cold fish is swallowed.

To determine stomach contents, it was once thought necessary to use an emetic, find freshly dead birds or take (that is, kill) specimens. The method of flushing out the bird's stomach with water is now used and is much more preferable as it leaves the bird unharmed.

Small pebbles are sometimes flushed from penguins' stomachs. The reason for swallowing them is not clear; perhaps they help grind up the food. The Emperor Penguin, the one least likely to come into contact with pebbles, dives deeply enough to glean them from the seabed or from the underside of icebergs that have ground them out of the rocks of the Antarctic continent. Other species have been noted picking them up from the seashore.

Most of these methods of study require handling the penguins, but there are some less invasive techniques. Coloured dyes squirted at wild penguins from a distance can be used for short-term identification (until the next moult). King, Emperor and Gentoo Penguins have red or orange on their bills, colours that may be used in recognition by the birds themselves, and it may therefore be better to avoid using these colours for marking purposes.

Another simple non-invasive method used to gauge how long a penguin has been ashore is examining its excreta. This changes from white (or pink if it feeds on krill) in freshly fed penguins through yellow to green after they have fasted for three days.

## DISPLAY: POSTURES AND CALLS
Penguins — like other animals (including humans) — exhibit ritualised behaviour. These rituals consist of postures, from head and

flipper waving to bowing and the presentation of gifts (nest material but not food), accompanied by calls. Rituals vary from species to species; some are complex, others simple. These rituals are described in later chapters for one species only in each genus, and in greatest detail for Adelies.

Varying interpretations are given to displays, ranging from the classical to the message-meaning to the game theories. In general, when birds do not recognise each other or dispute territorial claims, aggressive displays act as a warning: 'Get out and I mean it'. They are of varying intensity and are thought to influence an opponent's behaviour because it understands the intention of the bird displaying. A penguin that indicates it will attack is likely to repel an opponent more effectively than if it seems undecided whether to stay or flee. If the opponent does not take the hint, fighting often follows. This can be quite savage with chases, harsh squeals, bills interlocked, and birds falling to the ground and thrashing each other with their flippers, taking not the least care of whether they are trampling over nests and chicks including their own. In extreme cases, birds are killed in these fights.

When penguins are ready to breed, in some species at two years of age but in most at more than five years, they begin courtship by elaborate displays with postures and calls. The male, usually arriving first at the breeding area, advertises himself with some version of the Ecstatic Display which indicates 'I am the greatest and I'm available —try me'. When a pair begins its courtship ritual other pairs in the vicinity are likely to join in, a common occurrence where birds breed colonially. Pairs exposed to these rituals will court and copulate more often. This results in earlier and more synchronised laying, hatching and rearing. Having large numbers of chicks hatched at the same time makes it more difficult for predators to be effective — safety in numbers.

Penguin calls consist of harsh braying sounds, trumpeting, duetting, growls, trills and yelps, as the breath is inhaled and exhaled. When a pair calls together the calls are not in synchrony. Partners may display with short antiphonal calls (where calls are given by

one bird in response to the other). The African Penguin (as well as the spheniscids of South America) is often called the Jackass Penguin because it sounds like a donkey braying.

Displays are given so that birds may recognise each other, both as common species and as potential partners. Once a bond has been formed, partners continue to signal their intentions by sexual display. The display is repeated whenever birds meet again after one has been to sea and returns to the nest.

All these rituals bear some similarity to those employed by a man seeking a mate: advertising his intentions ('How about it?'), his territorial defence and aggression towards those with similar intentions ('Push off, you, this is my beat'), his bearing of gifts and bowing to the female (even if rather old-fashioned these days), and his attempt at song. It is up to the female to consent or reject. If rejected, the male penguin does not force his attentions on the female. If he tries to copulate and she is unwilling, she stands up and tips him off.

Though sleeping penguins are popularly depicted with heads tucked under flippers, there is little in the literature about their stance when asleep. Little and Rockhopper Penguins do not tuck their heads away. King Penguins do, their bodies partly supported by their tails, resting back on what look like their heels but are not. (Bird legs are unlike those of humans; what appears to be the knee is actually the equivalent of our heel.)

## BREEDING COLONIES

Penguins breed in colonies. Nests are either loosely scattered or tightly packed together, with territories extending only within pecking distance of the nest or the pair. Large closely-packed colonies are noisy and smelly, though chicks normally do not foul their own nests. Even those in burrows defecate into the area surrounding the nest. There is constant noise from greeting ceremonies or squabbles between neighbours. Occasionally penguins

Fig. 9. *Left* Royal Penguin braying as part of nest display.

will nest away from other penguins, but breeding is often unsuccessful when they do so. Yellow-eyed Penguins are different: they
always nest well away from each other. Colonies of penguins are
often called 'rookeries' but rookeries are really for rooks (a very
different bird, a member of the crow family).

Fig. 10. Royal Penguin breeding colony.

## BREEDING SEASON

Breeding is generally synchronised in Antarctic penguins (except
for Kings, which are also subantarctic), but in temperate climates
it can be spread over a long period. Specific seasons are given in
each chapter. Southern hemisphere seasons are regarded as: spring
— September to November; summer — December to February;
autumn (fall) March to May; winter — June to August.

## PAIR BONDS AND NESTS

Pair bonds are often long-lasting and many birds return to meet their partners at the same nest site each year. Kings and Emperors do not have nests, but they return to the same area to breed. During the time they are ashore, penguins do not eat. Nest-building is carried out by both males and females with whatever is available: plant material, seaweed, bones and stones. Stones not only stop the eggs from rolling away but also keep them out of the water when snow and ice melt. Surface-nesting colonial species steal stones and nesting material from each other, such thievery often resulting in fights.

## INCUBATION, HATCHING, CRÈCHING AND FLEDGING

Breeding strategies differ between the species. Those that produce two chicks may, in response to poor feeding conditions, feed the chick that begs hardest (facultative brood reduction), ensuring some limited breeding success. In some species siblings, as they grow older, have to chase their parents to be fed. This ensures that the first hatched, and therefore usually the larger, is fed first. In times of food shortage the older thrives and the younger dies. Other species discard one egg even though it is viable (obligate brood reduction).

Except in Emperor Penguins, both parents take it in turns to incubate the egg or eggs, usually starting at the laying of the second one. All penguins share the care of the young. In many species of bird the feathers are arranged in tracts along the body, but in penguins the body is covered completely except for a bare patch low on the abdomen, the brood patch. In breeding birds this brood patch becomes swollen and suffused with blood and transmits heat to the egg. Without the brood patch the bird's feathers would prevent the egg being warmed sufficiently for the embryo inside to grow. During incubation the bird usually lies prone over the egg or eggs, bringing them into contact with the brood patch. In the break between incubation shifts, the brood patch subsides and feathers come together to insulate the body and exclude water.

The chick calls from inside the egg when it is ready to hatch, cutting through the broader end with the egg tooth on the end of its bill. When it hatches the chick is covered with a thin coat of down and needs the warmth of its parents for the next two or three weeks, by which time it has grown a second, thicker downy coat.

Newly hatched chicks stretch their heads upwards, wave them around and give constant shrill piping calls. When older they peck towards their parents' bills, sitting up balanced by their feet and fat bellies. Penguins do not have crops. Instead, food is stored in the stomach and fed to the chicks by regurgitation; that is, the parent bends over and the chick thrusts its head into the parent's open mouth. When the chick is older the parent regurgitates straight into the chick's mouth. Only chicks are fed in this way.

Fig. 11. Gentoo Penguin feeding large young.

There is no transfer of food between members of a breeding pair during courtship, nor by any other penguin once it goes to sea to feed itself.

At three to four weeks old, chicks no longer need their parents to keep them warm and protected. The chicks of some species remain at their nests, but others abandon their nests and join together with other chicks to form a crèche. Adults do not guard crèches. Crèching takes place at the time that both parents go to sea to meet the increasing demands for food of their rapidly growing young. Chicks huddle together for warmth and safety against predators, which find it easier to pick off single chicks than those in big crèches.

Experiments and analyses of calls have shown that there are distinct individual differences in the calls of both chicks and parents. These differences enable parents to recognise their own young, which they feed exclusively, and chicks to recognise their parents. Most adults lead young crèching chicks back to the nest, if only briefly, before feeding them. When chicks are starving they may beg from any adult returning from sea, but without success.

Fully feathered chicks go to sea and swim without any instruction. 'Fledgling' is the term applied to a fully feathered young flying bird, and it is said to have 'fledged' when it flies from the nest. It is also an appropriate term for penguins going to sea for the first time because they actually do fly — under the water. Parents do not accompany the fledglings which therefore must be totally self-reliant in finding food. Juveniles disperse from the breeding area, some travelling vast distances in their first few years of life, before returning to their natal colony to breed.

BREEDING SUCCESS
Breeding success varies so widely that only a general idea will be given here. Unless conditions for each year are considered, 'average' figures mean little. Success depends largely on the amount of food taken by the adults, both before and during breeding and rearing of the chick, and on environmental factors such as

weather. In years when food is short, adults will be lighter in weight and therefore unable to withstand the long fasting periods ashore that some species must undergo. In some years they may abandon the attempt to breed at all. Some species breed for much longer periods when the food supply is good, going on to further clutches after successfully raising the first. Populations are never in equilibrium because there is variation in the quality of breeding seasons and in the frequencies of other adverse events: predators, both natural and unnatural; temperature anomalies; and oil spills, with their effect on food supplies. Consequently, breeding success is not generally mentioned in later chapters.

### MOULT

After their first year all penguins moult once a year (apart from Galapagos Penguins which may moult twice in a year). Worn feathers cease to be waterproof and thus do not insulate the birds' bodies from the cold. King and Galapagos Penguins have a pre-nuptial (before breeding) moult, but in other species moult follows the breeding season, with non-breeders moulting earlier than breeders. Before they moult, penguins spend from a few weeks to several months feeding at sea. When they have stored up sufficient energy to produce feathers and to maintain themselves during the moulting fast, the birds come ashore. The dull and ragged feathers no longer lie sleekly against their bodies and they open up, soon to be pushed out by the new ones underneath.

The stored fat and other reserves stored in muscle are released as energy. Physiological stress puts the birds at risk during moult and if they have been unsuccessful in storing up adequate reserves, their muscles deteriorate and they may die. Unlike flying birds, penguins are unable to feed during moult. If they went to sea they would become waterlogged and cold, so they stay out of the water for the two to four weeks it takes to moult. If starvation forces them to go back to sea to feed before moult is completed, they will die from waterlogging and cold. Those that survive the moult and grow new feathers are quite slim when they return to the sea.

Fig. 12. Moulting Little Penguin.

## CIRCADIAN RHYTHM — THE PENGUIN'S DAY

Penguins living south of 60° latitude experience more than three months of continuous daylight during summer, though there is an alteration in the intensity of light each day. During winter there is a corresponding period of continuous darkness, and between summer and winter they experience light and dark cycles of increasing or decreasing length. Yet they seem to adhere to a roughly 24-hour daily rhythm. It has been suggested that the Adelie in summer has a rhythm of slightly more than 24 hours, dictated by its internal clock. Both male and female penguins, whichever is not incubating, return to the nesting site around the time the eggs hatch. This ability is dictated by internal rhythm, perhaps initiated by hormones generated at the time of egg laying.

## HANDLING PENGUINS

Sometimes it is necessary to handle penguins, such as when attaching a flipper band or a device for recording movements or when collecting stomach contents. Some penguins stay in their burrows or stand their ground when approached by humans, so it is easy enough to pick them up, avoiding their beating flippers and biting bills. The larger or more timid species present a problem. A flying tackle is one method that achieves its end, but this is exhausting and can harm the bird (to say nothing of the field worker). Emperors are large and fast. Gentoos are also fast and can skitter along pebbly beaches and across lolling elephant seals. The pursuer sinks into the pebbles and is likely to be bitten by the elephant seals. A temporary corral provides one answer. Emperors and Gentoos can be herded slowly into the corral. Emperors follow the leader and are readily corralled. Gentoos present a greater problem as they flee approaching humans, except on Macquarie Island where they follow the example of the placid Kings. Inside the corral Gentoos try to evade capture, while Kings, which are usually moulting, look bemused.

## LONGEVITY

The larger penguin species do not breed until at least four years old, more often later. The medium-sized Royal Penguin, for instance, may delay breeding until 11 years old. All are long-lived and there are records of birds more than 20 years old (see individual species). This is not to suggest that most of them live to such an age. Moult is a time when they are vulnerable and many do not have sufficient resources to survive. These deaths are the most obvious; the majority of fatalities, however, probably take place at sea.

*Chapter 2*

# Genus *Aptenodytes*

INTRODUCTION

In the genus *Aptenodytes* (unwinged diver), there are two species: the King Penguin and the Emperor Penguin. They look alike, but the Emperor Penguin, the largest of all penguins, is less brightly coloured than the King Penguin. King Penguins breed on both sides of the Antarctic Polar Front, while Emperor Penguins breed only in Antarctica and are rarely found north of the pack-ice.

Both species walk in a sedate manner, planting their feet carefully and deliberately. They do not hop, but will toboggan on the ice and snow. Courtship behaviour does not vary much between the two, so it is described here for the King Penguin only.

Neither King nor Emperor Penguins make a nest. The single egg (and later the chick) is carried on the feet and covered with a loose belly fold which brings it into contact with the brood patch. They do not mate for life, the pair bonds lasting only for the one season, although some are renewed in succeeding seasons. Giant-petrels are the land predators most likely to harm their chicks which represent a regular food source for them.

There the likeness between King and Emperor Penguins ends. Their breeding distribution does not overlap and, even if it did, their chicks could never be mistaken for each other. Emperor Penguin chicks are all grey-white with some black on the head, whereas King Penguin chicks are brown all over.

# King Penguin *Aptenodytes patagonicus*

## DESCRIPTION

The specific name *patagonicus* refers to Patagonia, probably the region in which King Penguins were first described.

With a length of 85–95 cm, a flipper of 32–34 cm and a body weight of 12–14 kg, the King Penguin is the second largest penguin. It is a stately bird, with silvery grey-blue upperparts, a black tail and white underparts. The head, chin and throat are black. A bright golden ear-patch extends as a thin narrowing stripe around the sides of the neck to the upper breast. This is golden orange fading downwards to yellow, and merges with the white underparts. A thin black stripe separates the orange ear-patch from the blue-grey nape. It extends around the neck to the sides of the breast, broadens above the base of the flipper, then continues down to the flank. This stripe separates the white underparts from the blue-grey back.

The black bill with orange and pink plates is long and slender and slightly down-curved. The eyes are brown and the feet and legs black. The second down of the chick is thick and long and uniformly brown, resembling a long fur coat with a hood. The plumage of fledglings is similar to that of adults, but the ear-patches are pale yellow.

## DISTRIBUTION, DISPERSAL AND POPULATION

King Penguins breed on subantarctic and Antarctic islands between 46° and 55° south. At sea they forage north of the

Fig. 13. *Right* King Penguin colony (Macquarie Island).

pack-ice where Antarctic and subantarctic surface waters meet (the Antarctic Polar Front) in the southern Atlantic Ocean, southern Indian Ocean and the Australasian sector of the Southern Ocean. Breeding birds generally stay close to their breeding areas.

All year round, many birds are present in King Penguin colonies or in small parties on the beaches. Immature and non-breeding birds disperse and travel many kilometres from land, but none has so far been recorded in the South Pacific Ocean between Antipodes Island and South America. Stragglers have landed on the Antarctic continent, as well as on the coasts of New Zealand, Australia, South Africa and South America.

The population of King Penguins south of the Antarctic Polar Front was estimated in 1993 at 1.02 million breeding pairs (95 per cent of the world population).

## AT SEA AND ON LAND

At sea King Penguins group together only near the colony. Further offshore they form small groups or are solitary. King Penguins fitted with transponders (electronic devices) were found to dive deeply during the day (the deepest dive being 322 m), and the deep dives were found to be more efficient than the shallow. At night their prey rise towards the surface of the sea and then the dives do not exceed 30 m. Parents engaged in feeding chicks dived both day and night.

King Penguins have the largest span of any penguin between the tips of the open bill, presumably allowing them to take larger prey.

## BEHAVIOUR

The King Penguin seems quite unafraid of humans and often appears to be curious about what they are doing. They stroll unperturbed through human settlements and in semi-darkness can be mistaken for little people. On a beach on Macquarie Island, as I struggled to band a Gentoo Penguin, a King Penguin peered over my shoulder as if interested in what I was doing. Still curious, it walked in front of me, probed at my boots and accoutrements and, no longer interested, sauntered away.

King Penguins are highly gregarious at their breeding colonies, but the breeding adults are separated from those roosting, moulting or otherwise unoccupied with breeding. Even in the coldest weather, King Penguins do not huddle.

The King Penguin displays aggression and uneasiness by ear-rubbing and preening of its shoulder and the upper parts of the flippers. Wing-flapping is used to warn other birds. Horizontal head-circling, where it throws its head back, then moves it forward in a lateral half-circling motion while grunting, may denote the intention to strike with its bill. In the Direct Stare display, the bird faces its opponent with the flippers raised and the bill stretched out, an indication of low-level aggression. Fighting is rare, but birds may strike opponents with their flippers or pinch with their bills. In defence, when a bird is trying to reach its position among other birds, it walks with upstretched body, head high and flippers slightly raised, trying to avoid flipper blows and pecks.

When a male arrives in the colony to breed he advertises his presence by standing and calling with his head lifted, bill nearly vertical, neck fully extended, eyes half closed and flippers held slightly out from his body. His back is slightly concave and his tarsi (what looks like the lower leg) and feet are raised from the ground so that he appears to be leaning back on his heels. He stands still in the interval between displays. If he is successful in attracting the attention of a female, the two birds stand facing each other 1–2 m apart, shaking their heads vigorously, punctuated by short antiphonal calls. After this, one bird (usually the male) leads off in an Advertisement Walk, an exaggerated waddling gait with the head making a pendulum motion. It is followed by the partner.

When courtship has advanced, the two birds face each other and slowly lift their heads to stretch as tall as possible and hold the posture for about ten seconds, followed by head-shaking before relaxing. Later the birds engage in antiphonal duets and dabbling, in which one bird bends down and is imitated by its partner. Bill-clapping may take place during these bows. During one of the female's bows, the male hooks his neck over hers and presses downwards. The female slides to the ground whereupon the male mounts for copulation, steadying himself by pinching his mate's neck with his bill and pressing his flippers against her sides.

Fig. 14. King Penguins copulating.

At the change-over during incubation and brooding of the young, both birds indulge in mutual display, which involves antiphonal duets and dabbling. King Penguins allopreen.

## BREEDING

The classical explanation of the King Penguin breeding regime is that a pair will breed early one season, take more than a year to rear the young, and breed late the following season so, by the time the young has fledged, it is too late in the third season to breed again. Therefore, in the best of circumstances, they breed in two out of every three years.

Fig. 15.   King Penguin incubating.

More recent study in the Crozet Islands has shown that this explanation does not fit conditions there. The first season is the same as the classical model, that is the birds breed early in the season. Those that succeed in rearing young may lay again late the following season, after February, but invariably fail to raise young this second time. They recover condition quickly after this failure and are able to breed successfully in the third season. It has yet to be shown that this breeding regime applies to other King Penguin breeding colonies.

King Penguins generally do not associate with other penguin species, although when they are not breeding they mix with Gentoos in some places.

They form colonies on beaches, and in valleys and moraines free of snow and ice, preferably on level ground and at low altitude. They need easy access to the sea, but at times colonies may be 100 m above sea level and 1500 m from the sea. Colonies are very noisy during breeding with the loud trumpeting calls of adults, but in the winter the sound largely comes from chicks with their clear three-note whistles. Adults are territorial only from the end of courtship through incubation to the end of brooding. The birds shuffle only a few metres with the egg on their feet and their territory reaches only into that area within individual pecking distance.

Pair bonds are often not renewed annually because arrival at the colony is poorly synchronised. A pair that breeds successfully may break up because the female ceases to feed the chick a month earlier than the male. She may have finished her moult before his has begun and will go on to breed with another partner.

At any King Penguin colony all birds will be at a different stage of breeding. Eggs are laid from November to April and both parents share incubation of the egg for eight weeks and brooding of the chick for seven weeks, with the chick standing on the parents'

Plate 1
King Penguins.

feet. Sparse down covers the chick at hatching, to be replaced by thick brown down by the end of three weeks, with a facial mask and neck of light grey.

Chicks then form crèches, but even when it is very cold they do not huddle together. They are able to maintain an even temperature, unlike Emperor Penguins that do have to huddle. King Penguin parents feed their chicks in the crèche, but do not guard them. Chicks flee from predators into the crèche, and large chicks can successfully repel predators as large as giant-petrels by striking with their flippers and pecking.

During the winter chicks may not be fed for up to five months. Marine resources are too scarce to maintain both parents and chicks, so feeding is infrequent, if at all. Those chicks without sufficient bodily reserves do not survive.

Chicks remain in the crèche for nine months before fledging, by which time they have moulted their long brown downy coats and are fully feathered. They fledge alone and thereafter have no dependence on their parents.

King Penguins may begin to breed at the age of four years and will go ashore annually anywhere during those interim years to moult, sometimes on the shores of more temperate regions.

By the time the chicks fledge, the parents have been engaged in breeding (from courtship through to fledging of their young) for 14 or more months.

When moulting, King Penguins stand in groups on beaches while all their feathers are replaced over a period of 27–36 days. This is followed one or two months later by replacement of the coloured mandibular (bill) plates. To walk along a beach where King Penguins are moulting is like walking through a burst feather

Plate 2
**Emperor Penguin *Aptenodytes forsteri***
1. Adult, 2. Juvenile, 3. Downy young
**King Penguin *Aptenodytes patagonicus***
4. Adult, 5. Juvenile, 6. Downy young

Fig. 16. King Penguin crèche.

quilt. Beautiful patterns are formed where feathers float on stagnant pools.

### THREATS AND CONSERVATION

At sea King Penguins are taken by leopard seals and killer whales. On land skuas and sheathbills take eggs and small chicks, and older chicks are killed by giant-petrels. Human activities have impinged little on King Penguin colonies, although in the bad old sealing days they were boiled for oil.

# Emperor Penguin *Aptenodytes forsteri*

## DESCRIPTION

The species was named *forsteri* after the naturalist J. R. Forster who accompanied Captain Cook. The Emperor Penguin is a truly remarkable animal with its adaptation to breeding throughout the winter in the most hostile environment on earth.

With a length of 100–130 cm (a standing height of almost a metre), a flipper of 30–40 cm and a body weight of 30–38 kg, the adult Emperor Penguin is the largest of the penguins. It is very robust and its head appears small in relation to its body and feet. Its upperparts are blue-grey, its tail black and its underparts white with a pale yellow flush through the belly plumage. Its head, throat and chin are black and are clearly separated from the yellow-white upper breast. A broad yellow ear-patch, brightest near the head, fades to pale yellow along a broad stripe that leads to the upper breast. A black stripe extends upwards to separate the ear-patch from the upper breast, then downwards from the neck to the flipper to separate the upperparts from the underparts. Its slender down-curved bill is black with conspicuous orange, pink or lilac plates. Its iris is brown and its feet black. Young in thick, hairy second down are grey. The white face appears mask-like surrounded by the black chin, bill and forehead with an edging of black from above the eye to the side of the neck. Immatures are similar to the adults but duller in colouring.

Emperor Penguins are able to withstand the intense cold of the Antarctic winter mainly due to their plumage, which is extremely dense and tightly packed, and aided a little by thick deposits of fat on the body.

DISTRIBUTION, DISPERSAL AND POPULATION

Emperor Penguins live within the pack-ice of the Antarctic zone and generally avoid the open waters beyond it. They disperse from their colonies at the end of breeding in December or early January and return in March. Breeding colonies form only on the ice of the Antarctic continent and adjacent islands, with the exception of two colonies located wholly on land. These colonies form either on a frozen lake or on packed snow over rock and gravel.

The population in 1993 was estimated to be 195000 breeding pairs. In a smaller area, the Ross Sea coast of Antarctica where breeding sites are on annual sea-ice, it was about 60000 pairs in 1983.

When the ice clears each year it takes with it dead birds and eggs from the previous year, enabling adult and chick mortality as well as egg loss to be assessed. Another method of counting is to photograph a colony from above at different stages of breeding, using a balloon carrying a remote-controlled camera or from a helicopter. This provides information on the numbers of birds incubating and brooding and on crèched chicks.

AT SEA AND ON LAND

If the sea beyond the fast-ice is inaccessible for feeding, Emperor Penguins will dive into seal holes or cracks in the ice. Small fish and crustaceans are taken in shallow dives, larger fish and cephalopods in deeper dives. The deepest penguin dive recorded, more than 400 m, was made by an Emperor Penguin. Emperor Penguins have been known to stay under water for 18 minutes, though the duration of a dive is usually much less.

BEHAVIOUR

The displays of the Emperor Penguin, both aggressive and sexual, are similar to those of the King Penguin, but differ a little during courtship. On arrival an unpaired male stops moving through the colony, lets his head fall onto his chest, inhales deeply and utters a call with his head lowered. He may repeat this several times at short intervals. In response, a receptive female stands face to face with him and both slowly lift their heads, stretch as tall as possible, freeze for several minutes, then relax, either to part or stay together as a pair. Emperor Penguins do not allopreen.

## BREEDING

Most Emperor Penguins never walk on land. They leave the sea in March and trudge across the ice in long lines for up to 200 km to reach their breeding colonies. These are on the sea-ice that forms annually around the Antarctic continent, or sometimes on hard packed snow on easily accessible glaciers. Because territories exist only in the immediate vicinity of a pair, mates from previous years cannot spend time in searching to renew bonds. Pair bonds form quickly and there is no pre-egg-laying exodus as with many species.

Fig. 17. Emperor Penguins trudging.

In May, about six weeks after arrival, the female lays her egg. Egg-laying is highly synchronised throughout the colony. As soon as the egg is laid, the male scoops it up with his bill from between the female's feet and places it on his feet, where it comes into contact with the brood patch and is covered by the abdominal fold. He rests back on his tarsi ('heels') so that his feet make the least contact with the ice. The female is now free to set off across the ice to the sea, where she will feed.

It is now winter and is dark all day and night. Temperatures drop below −60°C and blizzards driven by winds of up to 180 km/h coat the birds with snow. The ability of male Emperor Penguins to withstand the intense cold of the Antarctic winter is due not only to their dense plumage, but also to the practice of huddling. When there is a blizzard the birds huddle together for warmth. This huddling behaviour is unusual because most other species tend to stay pecking distance or more apart, except within a family group.

The huddle is in slow but constant movement as those Emperor Penguins with their backs to the blizzard shuffle to the front of the pack in an attempt to get into the middle of the huddle where it is warmest. In dense huddles there may be ten adult males to a square metre.

Even though only a small amount of excreta is voided by individuals, the output of the thousands of incubating males causes the ice on which they stand to become stained dark green.

After nine weeks of incubation the egg hatches in July. Despite all this time ashore without feeding, the male is able to give the newly hatched chick a small feed.

The female returns soon after the egg hatches and the pair indulge in noisy greeting. The male needs some persuasion to let the female take the chick onto her feet, where she covers it with the abdominal fold. If during its transfer from the male's to the female's feet the chick slips onto the ice, it will freeze quite quickly or may be stolen by an unmated or failed breeder. Either way its chances of survival are slight because the intruder will eventually abandon it without a mate to relieve it.

At last the male is able to set out for the sea to feed again. After fasting for at least 15 weeks he has lost about half his weight. Even if the ice has begun to break up he still has a walk of about 100 km to reach the open sea, though he can enter the water through tide cracks or through seal holes in the ice. Altogether his fast may have lasted six months.

He has passed through three phases of loss of weight: an initial rapid loss, then a slow steady drop, followed by a final rapid loss. At one study site observed for 30 years, no adult male Emperor Penguin was found to have died of starvation during his long incubation fast. If his physical condition is inadequate to continue his fast, it seems that an internal signal stimulates him to abandon his egg and return to the sea to feed. If the female fails to return on time he abandons the newly hatched chick.

With the chicks requiring large amounts of food, two long lines of Emperor Penguins are in continual motion as they travel to and from the open sea.

Parents take it in turns to look after their chicks until, at about seven weeks old, the chicks join a crèche and huddle together for warmth and protection, unlike King Penguin chicks that never huddle. Parents continue to feed their chicks until December, locating them in the crèches by their high-pitched squealing calls. The chicks must also remember their parents' calls and answer when they return with food.

Chicks fledging from Taylor Glacier have to travel about 60 km to reach the ice edge. They assemble in groups to leave the colony during the day in the warmest weather, many of them down-covered but with feathers fully formed beneath. Chicks, like adults, follow each other. By their movements and calls, other chicks, though not quite ready to fledge, are induced to join in the trip to the edge of the sea-ice. Early in the season of departure they follow adults that are still feeding young. These adults take the shortest route to reach a polynya (open water surrounded by ice). They appear to be able to navigate by the reflection of the water on the clouds, a 'water sky'. Later fledging chicks, now abandoned by adults, also follow the shortest route. Previously marked by the constant passage of feet, this route is now obliterated by wind and snow, so presumably they navigate instinctively by the 'water sky'. Although this appears to be an explanation of how Emperor Penguins navigate *away* from the colony, there is no explanation of how they first navigate *to* their breeding colonies so far from the ice edge, in conditions of low temperature, decreasing daylight and blizzards.

Emperor Penguins do not breed until at least four years old and they may live to be 20 or more years old. They take a month or more to moult in January and February, standing around on fast or floating ice or on Antarctic shores. Having regained condition after moult, they again begin the cycle of breeding by trudging across the ice to their breeding colonies.

## THREATS AND CONSERVATION

When the ice does not break up, adults have to travel too far to forage and chicks in crèches die of starvation. In mid-summer it is possible that competition from commercial fishing may be to the detriment of Emperor Penguins during the important fattening phase that leads to the moult. The advent of humans on the Antarctic continent, however, does not appear to have upset Emperor Penguin populations. Leopard seals and killer whales prey on Emperor Penguins at sea.

The Taylor Glacier colony, 95 km west of Mawson and located wholly on land, has been declared the first Specially Protected Area under the Antarctic Treaty.

*Chapter 3*

# Genus *Pygoscelis*

## INTRODUCTION

The genus *Pygoscelis* (elbow legs, a description of little meaning) is composed of three species: Gentoo, Adelie and Chinstrap Penguins. They belong to the group of long stiff-tailed penguins (pygoscelids) whose tails sweep the ground behind them as they walk, as well as being used as props. Adelie and Chinstrap Penguins are almost the same size; both are a little smaller than Gentoo Penguins, which are smaller only than Emperor and King Penguins.

The breeding distribution of all three species overlaps only on or near the Antarctic Peninsula. During winter Adelie Penguins stay in the pack-ice zone in the vicinity of the continental shelf, while Chinstrap Penguins disperse north of the pack-ice. Gentoo Penguins generally remain all year round near their breeding islands and feed nearby.

Chinstrap Penguins are by far the most numerous of the pygoscelids, some three times more numerous than Adelie Penguins. Gentoo Penguin populations are small by comparison.

The dives of Gentoo Penguins are either short and shallow or long and deep. Adelie Penguins are capable of similar dives, but most are shallow. Chinstrap Penguin dives are generally short and shallow, but they have been recorded to dive the deepest of the three species.

Courtship behaviour is ritualised and is described in detail for Adelie Penguins. Allopreening has not been recorded for the three species nor is there any record of external parasites.

At South Shetland Islands, which all three species vacate when not breeding, it has been shown that the male Adelie Penguin is highly faithful to the nesting territory, but the pair bond is the least durable (the weakest among long-lived seabirds). Gentoo Penguins frequently change their nest sites, but their pair bonds are usually long-lasting. They are present all year at their more northerly breeding sites and are absent only intermittently during July and August from their sites in the Antarctic zone. Thus they are able to maintain their pair bonds at their nesting colonies. Chinstrap Penguins exhibit fidelity both to site and mate. It is suggested that the migrant Adelie and Chinstrap Penguins reunite at the nest site.

Males arrive first at the colony, select the site and start building before the females arrive. Together they build the nest on ice-free ground. Nests of Adelie Penguins are close together; those

Fig. 18.   Gentoo Penguin incubating while its mate adds nesting material.

of Chinstrap Penguins are slightly further apart; and Gentoo Penguin nests are separated by more than pecking distance. The two eggs laid are incubated for about five weeks. Chinstrap Penguins' eggs are of similar size, the chicks hatch the same day and are treated equally. Adelie and Gentoo Penguins differ from the Chinstrap Penguin in that the first laid egg is slightly larger and it hatches a day or two earlier than the second. When feeding conditions are poor the first chick is more likely to survive, especially when the chicks have to chase their parents to be fed. When parents are rearing a single chick there is no need for a feeding chase.

The breeding strategies of the three species may have evolved under different environmental conditions. Adelie Penguins are migratory and undergo long fasts; Gentoo Penguins are non-

Fig. 19. Gentoo Penguins in feeding chase.

migratory and do not fast except when moulting; Chinstrap Penguins are migratory, undergoing fasts somewhere between the other two species in duration.

The downy plumage of the chicks is different for each of the three species. Chinstrap and Adelie young take 7–8 weeks to fledge, while Gentoo Penguins take about 14 weeks.

# Gentoo Penguin *Pygoscelis papua*

DESCRIPTION

The specific name *papua* refers to Papua New Guinea, a tropical area entirely unsuitable for penguins. This is a result of the type specimen being mistakenly labelled as to its provenance.

Adult Gentoo Penguins are 75 cm long, with a flipper of 19–25 cm and a weight of 5–5.5 kg. The back and throat are bluish-black and the underparts white. A conspicuous white patch above the eye usually meets across the crown and white feathers are scattered across the head and nape. The iris is brown and the eyelid white, giving the effect of a white eye-ring. The bill is black and orange and the feet pale yellow to deep orange. The chick's second down is thick and hairy, dark brown on the back and white underneath. Fledglings have a duller bill than adults, the white bar over the crown is less conspicuous and most lack the white eyelids.

DISTRIBUTION, DISPERSAL AND POPULATION

Gentoo Penguins have a circumpolar distribution and breed on subantarctic islands and the Antarctic Peninsula. They generally remain near their breeding islands all year round, but stragglers reach into the temperate zone of Tasmania and New Zealand.

The bulk of the total world population, estimated in 1989 at 300 000–350 000 pairs, is situated near the Antarctic Polar Front where there is the greatest area of continental shelf available for feeding. The population that breeds south of the Antarctic Polar Front was estimated in 1993 at 236 000 pairs, 75 per cent of the world population.

## AT SEA AND ON LAND

Gentoo Penguins feed during the day close to the shore near their breeding areas. Their dives are either short and shallow or long and deep (up to 70–80 m and 2.5 minutes long). In southern colonies adult birds are absent for 5–6 hours during chick brooding, some leaving early in the morning and returning for others to leave in the afternoon. Most birds are ashore by evening. In more northern colonies, foraging trips are longer, often overnight during the period of chick-rearing when food demand is highest. Those that stay at sea overnight dive less frequently than during the day.

Birds walk into the water until their flippers are half submerged, then swim under water a short distance to emerge and bathe on the surface. On their return to land they porpoise to within a short distance of the shore, then swim on the surface to survey the area before coming ashore quickly on a wave. Gentoo Penguins are opportunistic feeders, with a great variation in the ratio of foods taken. In some areas fish predominate, in others krill. Females tend to feed on krill rather than fish, whereas males may switch from one to the other.

On icy coasts Gentoo Penguins leap out of the water onto ice in similar fashion to Adelie and Emperor Penguins, or jump onto rock shelves. They walk, run or toboggan, sometimes on sand, at speeds of 1.5–3.4 km/h and will also hop or jump over obstacles.

## BEHAVIOUR

Gentoo Penguins are probably gregarious at sea and breed in loose colonies on land. Ritualised threat postures and calls are used to convey aggression and courtship. Fights are common, but aggressive interaction with other species is avoided. In most places Gentoo Penguins are timid and will leave their nests unattended when approached by humans, but they may charge at predators. At South Georgia Island they react differently: adults often have to be lifted off nests by field workers collecting information on eggs and chicks.

Gentoo Penguins sleep lying prone with their flippers tucked underneath their bodies or at their sides, sometimes with the bill tucked under a flipper.

## BREEDING

The breeding season varies according to the locality, the colony and the year. There may be three months difference between the different localities: June (winter) breeding on islands near the Antarctic Polar Front and late spring breeding in the Antarctic zone. There is even a difference of several weeks in laying dates between colonies on the same islands. At higher (colder) latitudes, laying is more synchronous than at lower latitudes where sea temperatures are warmer. The earlier the laying, the longer the laying period, with the possibility of re-laying after an early failure.

At Crozet Islands (46°S), winter breeding is possible because Gentoo Penguins are opportunistic feeders, the winters are mild and there is no overlap with other penguin species and, therefore, no competition for food. At this time of minimum marine productivity, only low population levels can be sustained. At South Georgia Island (54°15'S), there is a greater period of penguin absence and breeding is limited to a short period in summer during the peak of krill production.

Inland colonies are situated on flat land. Coastal nest sites need to be away from floods and from elephant seals that crush the nests. At Macquarie Island sites are also situated on steep coastal slopes. At South Georgia nest sites appear to be stable, but at other places they change from year to year. Even with the change of nest site, most pair bonds are renewed annually, though in one year at South Georgia all previous pair bonds were broken. Both male and female birds are at the nest site well before breeding begins and they alternate incubation shifts daily, so that neither of the partners needs to fast for long periods.

The nest is a scraped out hollow lined with grass, stones, feathers, shells, bones and moss to form a mound. If there is no vegetation the birds will build with pebbles, but where possible nests are among tussocks and are sited more than a metre apart. At the end of the breeding season all the vegetation has been destroyed by the birds, with the result that nest sites are seldom used for two years in succession.

Chicks are brooded until they are aged three to four weeks, when they form small crèches in the nesting area. The age at which they crèche may be dictated by the parents: when food is in short supply both parents need to forage, this daily desertion causing the chicks to crèche. Small crèches join together and move onto the beaches to await the return of the adults. Immediately a parent of two chicks comes ashore, it is pursued by them in a hectic feeding chase which has the parent running in between and over the elephant seals lying on the beaches. Chases may take the birds 300 m from the colony. This separates the siblings and ensures that the larger is fed. When there is a food shortage the smaller chick starves.

Crèches become more mobile as time goes on and are often well away from the nesting site by the time the chicks fledge at about 14 weeks. This is almost twice as long as the time taken by Adelie and Chinstrap Penguins. Even then, fledglings may enter the water for only a short time before returning to shore and may continue to be fed for several more weeks. They do not disperse away from their own islands and are not migratory. When breeding conditions are bad, young Gentoo Penguins may delay breeding for the first time until they are three years old. When conditions are good, half the breeding young are two years old.

Adults begin to moult in January over a period of 15–21 days at their breeding islands. A temporary partial retention of body feathers (not recorded for other penguins) may possibly provide

Plate 3
**Chinstrap Penguin** *Pygoscelis antarctica*
1. Adult, 2. Juvenile, 3. Downy young, mesoptile, 4. Downy young, protoptile
**Adelie Penguin** *Pygoscelis adeliae*
5. Adult, 6. Juvenile, 7. Downy young, mesoptile, 8. Downy young, protoptile

Fig. 20.   Gentoo Penguin with chick.

Plate 4
**Yellow-eyed Penguin *Megadyptes antipodes***
1. Adult, feet flushed, 2. Juvenile, 3. Downy young, mesoptile
**Gentoo Penguin *Pygoscelis papua***
4. Adult, 5. Juvenile, 6. Downy young, mesoptile

Fig. 21.   Moulting Gentoo Penguins.

insulation for the duration of a short swim, allowing timid Gentoo Penguins to take briefly to the water to escape danger. Tail feathers are the last to grow, so post-moulting birds can be distinguished from pre-moulting birds.

### THREATS AND CONSERVATION

At unprotected nests on the edges of colonies, skuas harass nesting pairs of Gentoo Penguins. One skua struts in front of the nest until the sitting bird lunges at it in defence, whereupon a second skua seizes the egg or chick. Elephant seals flatten nests and leopard seals feed on Gentoo Penguins in the water. Kelp gulls, giant-petrels, sheathbills, caracaras (falcons) and feral cats also take young.

Because Gentoo Penguins are so wary of humans, they should not be approached closely unless it is necessary.

# Adelie Penguin *Pygoscelis adeliae*

DESCRIPTION
Adelie Penguins (and Terre Adélie in Antarctica) were named after the wife of the French explorer Admiral Dumont d'Urville, after whom part of Antarctica is named.

Adelie Penguins are most often represented as the prototype of penguins in advertisements and picture books. Adults are 70 cm long, with a flipper of 16–18.5 cm and a weight of 3.7–4.0 kg. Adelie Penguins are black above and white below, with a black face and a distinctive white eye-ring around the brown eye. In some displays Adelie Penguins raise the black feathers on the back of the head to form the occipital (back of the head) crest. The white feet and the short black bills, dull orange at the base, are almost covered by feathers for protection against the cold. At about ten days the young attain their second plumage which is thick, woolly dark grey down. Fledglings have white throats and lack the white eye-ring of the adult.

DISTRIBUTION, DISPERSAL AND POPULATION
Adelie Penguins are distributed all around the pack-ice of the Antarctic zone. They breed on the coast of the Antarctic continent and on islands where the land is ice-free and accessible from the sea. In winter they remain within the pack-ice zone, but stragglers reach subantarctic islands and even New Zealand and Australia.

The breeding population of Adelie Penguins was estimated in 1993 at 2.47 million pairs.

## AT SEA AND ON LAND

Using satellite tracking it was found that Adelie Penguins breeding near Mawson travel to broken ice as far away as the continental shelf in the pack-ice zone, wherever food is available. The distance to the ice edge decreases progressively through the summer until colonies in many places are bordered by open water.

On short trips Adelie Penguins mainly eat fish and amphipods, but on longer trips, mainly krill. Of ten birds tracked during the incubation period, females were absent for about three weeks (some of them for up to 36 days — too long a period for the male to continue incubation), while males were absent for shorter periods. Two female parents, off duty immediately after egg-laying, travelled 341 km and 243 km respectively from their breeding colony, making the return trip more quickly than the outward swim. The 341 km return journey took six days of swimming against ocean currents and a day and a half of walking over ice, giving an average speed of 2.4 km/h.

Though parents may stay away for long periods during incubation, when feeding young the period is short and they stay within 50 km of the colony. During chick-rearing the average swimming speed was estimated to be 4.6 km/h for periods of one and a half days. In bouts of diving during the day birds make about 50 dives in an hour, then have a half hour interval on the surface.

After moult Adelie Penguins continued to forage in the same area as when breeding.

Transmitters on breeding Adelie Penguins in 1986–87 showed that birds from Ross Island tended to concentrate for feeding in the middle of McMurdo Sound.

Adelie Penguins swim at speed under water or porpoise fast (25 km/h) up to the edge of the ice shelf, which may be up to 2 m high. They then leap out of the water, gripping the surface of the ice with their very long toenails. When leaving, they dive off the ice shelf after jostling at the water's edge; lurking leopard seals waiting for a feast are possibly the reason for this caution. If the rocky shore slopes steeply the birds flop into the water.

When travelling to breeding colonies, which are well inland, they walk or else toboggan at nearly 4 km/h, which requires less energy.

## BEHAVIOUR

Unlike the timid Gentoo Penguin, Adelie Penguins stand their ground when approached by humans. They are also aggressive towards skuas and will destroy skua eggs and chicks. This is not predation since they do not eat their victims, unlike the skua practice of predation on many bird species.

Both sexes defend territories using pecks and threats. Birds from adjacent nests interlock their bills, push and twist, and butt their breasts, the loser usually being the first to be knocked down. There are three main threat displays, all given by adults and un-defended chicks standing erect or prone with the flippers held by the sides, the eyes down, the occipital crest erect (Figure 22), and accompanied by calls. In the Direct Stare (Figure 23) the bird points its bill at the threat, such as an overflying skua; in the Fixed One-sided Stare (Figure 24) the neck is arched and one side

Fig. 22.  Crest erect, eyes down.          Fig. 23.  Direct Stare.

Fig. 24.  Fixed One-sided Stare.

57

of the face is presented perpendicular to the threat; and in the Alternate Stare (Figure 25) the crouched bird arches its neck and twists its head back and forth to show alternate sides of the face, accompanied by a slow flipper wave.

Fig. 25. Alternate Stare.

In the Crouch (Figure 26), the bird moves to place its centre of gravity over its deeply flexed legs, the head and neck are withdrawn and the bird gapes and pecks or calls.

Fig. 26. Crouch.

Fighting, common before laying and after re-occupation of the site (but not between mates), usually begins with the breast thrust forward. Then the birds bump each other and beat rapidly with their flippers while attempting to hold onto each other with their bills. The fight ends when one of the pair falls over, scrambles up and runs away, often through the colony, with both pursued and pursuer being pecked at by nesting birds (Figure 27). Chicks fight similarly.

Fig. 27. Fighting.

When a penguin stranger lingers near the nest or a skua approaches, Adelie Penguins will run a few steps forward and charge with upright bodies, raised occipital crests and bills pointed forward. Pecks are a mild form of attack aimed at birds passing an occupied nest or at a neighbour. They are common throughout the season but mostly during nest-building. If a female is aggressive, moves quickly or does not bow her head, the male often pecks her or drives her away. When they are among nests, birds hold their feathers tightly against their bodies, necks elongated and flippers held back, in an attempt to reduce body size and prevent aggression.

There are numerous displays in pair formation, most accompanied by various calls, but the following are the main postures adopted by Adelie Penguins.

The Ecstatic Display starts with slow rhythmic flipper beats as the breast is thrust out and the neck arched. The entire body is then elongated and pointed upwards. As the climax is reached, the neck and bill are stretched to the maximum, the sides of the neck expanded exposing the pattern of the throat, the eyes are rolled down and backwards, the occipital crest is raised, and the flippers are beaten horizontally (Figure 28). This display is common in males and is used to attract a female to the nest, but is rare in females. After the climax, the bird enters the Bill-to-axilla Display (Figure 29). It leans forward, rocks and twists its head from side to side, and directs or tucks its bill underneath either flipper and away from the recipient, the flippers either beating or motionless.

Fig. 28.   Ecstatic Display.          Fig. 29.  Bill-to-axilla Display.

In Bowing, the members of a pair arch the head and neck forward and point the bill at the ground (Figure 30) in a deep bow, with the flippers held tightly to the sides and slightly forward.

Fig. 30. Bowing.

The Oblique Stare Bow is given by a bird in the upright posture lowering its head to below horizontal and presenting one side of its face towards a bird of opposite sex, with the bill closed. Males sometimes have eyes down, but females do not. The 'crest erect' is always shown by the male, but the female relaxes or only partially erects her crest. The male's bill is pointed much more towards the female than hers is to him.

In the Loud Mutual Display, commonly given by paired birds, the two face each other, bills pointed upwards and open wide, necks stretched out, eyes rolled down and back, occipital crests raised, flippers held to the sides (Figure 31) and the heads, necks and sometimes the bodies are swayed from side to side as a distinctive call is emitted. In the Quiet Mutual Display, movements are

Fig. 31. Loud Mutual Display.

less pronounced, the bills are closed and the birds are either quiet or call softly. Either display is seen after disturbance in the colony or in response to the Ecstatic Display and often following Mutual Bowing.

In copulation, the male bows low to the female as she lies on the nest and then approaches her from the side in the Arms Act. He may mount with few preliminary movements or after a prolonged period of bowing. With his head bent low and flippers beating he mounts the female half-way along her back; the female raises her bill and the male's bill vibrates when it contacts her bill and chin. Treading on her back, the male moves backwards, tail wagging from side to side against her upturned tail and there is rapid apposition of their protruding cloacas as the male reaches the female's rump (Figure 32). The male then jumps off, walks around to the side of the female with his head bowed, and the female flicks her tail forward in jerky movements. Mounting is repeated many times during the pre-egg period.

Fig. 32. Copulation.

Displays are associated with the nest, but they lessen after the eggs hatch and the chicks grow larger, and stop when the chicks enter the crèche.

Sleeping Adelie Penguins either lie down or stand up, with the bill horizontal and head withdrawn to the shoulders. Occasionally, they will stand with the bill tucked to one side, sometimes behind the flipper.

## BREEDING

Adelie Penguins must complete their whole breeding cycle in the short period between October and February. The days are long, in fact from November to January daylight is continuous in most of Antarctica. When they arrive at their breeding colonies they may have travelled up to 100 km from the nearest open water. Colonies may contain up to half a million birds and are situated on rocky or pebbly places, on flat or irregular terrain, on beaches or on hill slopes, but never on ice. Sites are exposed to the sun and often to wind, both of which keep them free from snow drifts. Adelie Penguins need access to open water near the pack-ice for feeding.

The nest site is the focus for reunion. If both members of a pair do not arrive at much the same time, rather than miss out on breeding, the birds tend to pair up with any other willing penguin. As a result, pair fidelity is the lowest of all the long-lived seabirds. Lack of fidelity may not be so much by choice (divorce) as by death of the partner, as Adelie Penguins suffer a high mortality (see 'Threats and conservation'). It is the female that makes the choice of partner. If her previous partner already has a new mate, she will try to drive her away. If she has already mated and her old mate arrives, she will desert her new mate for the old one.

Adelie nests are placed just beyond pecking distance of their neighbours. The nest scoop or hollow is lined and surrounded with pebbles up to egg size and sometimes with bones. Additions are made by both birds during nest relief with stones gathered and often stolen from neighbours or from distant unoccupied nests. Thievery is constant.

Laying starts 12 days after the arrival of the first birds and all two-egg clutches are laid over one month in November/early December. By this time both members of the pair have fasted for about three weeks, though females often leave the colony during this period to make short trips to the sea-ice to eat snow or ice. The male takes the first incubation shift. When he arrives in the colony he is heavier than the female. This stored energy is necessary to sustain him until the female returns to take over her shift, by which time he will have fasted for four to six weeks. Adelie Penguins hold their eggs on top of their feet, somewhat like Emperor and King Penguins, but also lie over them in a manner similar to other penguins.

When the chicks' second down starts to be replaced by feathers at about three weeks, the chicks crèche; small crèches amalgamate into larger ones. Chicks leave the colony at about seven to eight weeks, independent of their parents, but they may take short swims for two or three days before departure.

The earliest age of first breeding for females is three years and for males four, but usually they do not breed until one or two years later.

Fig. 33. Adelie Penguin and chick.

Breeding success or failure depends on the availability of food, both the quantity and possibly the distance of travel necessary to obtain it. This can lead to the failure to co-ordinate nest relief. When pack-ice necessitates too long a journey to open water, either the mate will desert after too long a fast or the chicks will starve. Hatching success relates to the time of ice break-out; it is poorer when the break-out is later.

Adults take about three weeks to moult ashore or on the sea-ice or on icebergs, sheltering in the lee along with adolescents, though some adolescents will moult on the mainland.

## THREATS AND CONSERVATION

Mortality among Adelie Penguins is high. Adults are at risk when breeding, especially the young adults. In one study it was shown that 75 per cent of three-year-old and 40 per cent of four-year-old breeders did not survive the breeding season. To rear their young, breeders must run the gamut of leopard seals 47 times in going to and returning from the sea. As several leopard seals line the departure beach waiting for a feed, each journey is fraught with danger.

Adelie Penguin breeding success is adversely affected by a colony's proximity to human habitation and disturbance; their numbers decline and skuas, which tolerate humans, prey on eggs and brooded chicks. Nests on the periphery of the colony are more exposed to predation by skuas.

During the chick-rearing period the foraging range of Adelie Penguins overlaps with areas fished commercially. There is, therefore, the potential risk of competition which would undoubtedly be detrimental to Adelie Penguins.

# Chinstrap Penguin *Pygoscelis antarctica*

## DESCRIPTION

The specific name *antarctica* implies locality, but Chinstrap is a more descriptive name.

Adult Chinstrap Penguins have a length of 71–76 cm, a flipper of 17–20 cm and a weight of 3.9–4.4 kg. They are black above and white below. A narrow band of black feathers, the chinstrap, extends from ear to ear, separating the white chin and cheeks from the white throat. The black bill is bare of feathers, the iris red-brown and the legs and feet are fleshy pink. The second down of the chick, which appears when it is about 20 days old, is grey on the back and face and pale grey below. The fledgling is similar to the adult but has some dark spotting on the face concentrated around the eye.

## DISTRIBUTION, DISPERSAL AND POPULATION

Breeding colonies are south of the Antarctic Polar Front and mainly in the South Atlantic and on the Antarctic Peninsula. The range of the Chinstrap Penguin that is usually stated includes occasional breeding attempts at Heard Island, but no successful breeding has been recorded there. Stragglers occasionally visit and may attempt to copulate with Gentoo Penguins. From March to early May, Chinstrap Penguins move well north of the pack-ice for the winter, even approaching the Antarctic Polar Front during August. Some stragglers, often moulting, reach southern Australia and South America.

Fig. 34. *Right* Chinstrap Penguin.

The breeding population of Chinstrap Penguins was estimated in 1993 at 7.49 million pairs.

## AT SEA AND ON LAND

When brooding young, Chinstrap Penguins forage for half a day at a time between 3 and 20 km from their colonies; they forage at greater distances during the remainder of the breeding period. They feed on dense concentrations of krill, with a few fish. Their diving effort is almost continuous and is concentrated at midday, although there is a second peak of activity around midnight. Most dives last about a minute and are within 50 m of the surface, but they have been recorded up to 121 m in depth and three minutes duration.

Chinstrap Penguins waddle on land or, falling down, they toboggan by propelling themselves by their feet without the use of flippers. All four limbs are used in climbing. Leaping more than a metre to slippery footholds, they crowd onto icebergs in such numbers that those already there are pushed off.

## BEHAVIOUR

Sexual display is similar to that of Adelie Penguins except that both sexes of Chinstrap Penguin perform the Ecstatic Display. They are gregarious away from their nests and are generally regarded as the boldest, most pugnacious and agile of the pygoscelid penguins. They will charge at an intruder rather than flee (as most Gentoo Penguins do), or stand their ground (as Adelie Penguins do). Alone among penguins, they can outmanoeuvre and will deliberately attack a dog. Humans entering a colony may be attacked by birds rushing at them, jumping and grabbing clothing. At sea they have been known to jump into open boats and rest there quietly.

Chinstrap Penguin aggressive and sexual behaviour is similar to that of the Adelie, but the ecstatic call is louder and more ear-splitting, giving rise to a deafening cacophony both day and night.

**BREEDING**

Chinstrap Penguins return in early November to their breeding colonies, some of which contain thousands of breeding pairs. Where they are mixed with Adelie Penguins, Chinstrap Penguins outnumber them and compete successfully in conflicts for nest sites, sometimes nesting at a higher altitude. Pair bonds are usually retained from year to year, but the nest site may change slightly. Sites are on ice-free rocky ground on headlands and foreshores, sometimes with sparse vegetation, or on raised terraces or cliffs up to 75 m above sea level. Nests may be built 90 m from the landing beach, the higher elevations being the first to be settled. Nests are built up with feathers, stones and bones, and are large enough to act as windbreaks. Stones collected by both parents are added during the incubation and nesting periods. Nests are about 60 cm apart, not as close as Adelie Penguin nests but not as far apart as those of Gentoo Penguins. Territory is restricted to the area around the nest.

Both parents lie over the eggs during incubation and most pairs rest or sleep on the nest, either prone or upright. It is possible that Chinstrap Penguins can defer incubation until the second egg is laid because their laying season is later than Adelie Penguins and conditions are not so harsh. The female arrives five days after the male and takes the first incubation shift. This means that the male, which is heavier than the female on arrival, does not have to fast much longer than three weeks, unlike the male Adelie Penguin, which takes the first shift.

Chicks do not crèche, but stay at the nest until they fledge at seven to eight weeks. They do not indulge in feeding chases, both chicks being treated equally by the parents. At first, fledglings stay near the colony. They may swim frantically, but do not dive and soon land again. All depart by early March. They first breed at the age of three.

Breeding success depends on the availability of food; persistent pack-ice restricts access to the sea for foraging.

Most birds moult on ice floes or icebergs in the pack-ice, where they shelter in the lee of hummocks or ridges, though some birds may moult ashore. The duration of moult is not well known, but may be as short as 13 days in late February.

## THREATS AND CONSERVATION
Predation of eggs and chicks is mainly by sheathbills and skuas and, at sea, by leopard seals. The effect of human disturbance on Chinstrap Penguin colonies is the same as for Adelie Penguins.

*Chapter 4*

# Genus *Eudyptes*

INTRODUCTION

There are six species in the *Eudyptes* genus (crested penguins): Rockhopper, Fiordland, Snares, Erect-crested, Macaroni and Royal Penguins, though the last two are often regarded as the same species. Three sub-species of Rockhopper Penguins are recognised but there is now a move to have them separated into two distinct species — the Northern and Southern Rockhopper Penguins — because there are differences in behaviour, song and bill measurements. For the purposes of this book Rockhopper Penguins are treated as a single species.

All the eudyptids have red or red-brown eyes and crests of yellow or orange plumes, some of them drooping, others erect or mixed with the blue-black feathers of the crown; and all have some degree of dark markings and lines on the underflipper.

Rockhopper, Macaroni and Royal Penguins live on Antarctic and subantarctic islands; Fiordland Penguins live in the temperate regions of New Zealand; and Snares and Erect-crested Penguins live on islands just to the south of New Zealand. Wherever two species of eudyptid co-exist, one is always larger in size than the other.

All eudyptids swim with their heads and parts of their backs above water, and porpoise when travelling fast. Except for Fiordland Penguins, eudyptids are migratory and may spend up to five months at sea when they are not breeding. There is little difference in displays between all six species and their calls are similar. The display of the Rockhopper Penguin is given as an example.

Allopreening is common between pairs, unattached birds, and parents and chicks.

All eudyptids are spring breeders apart from Fiordland Penguins which breed in winter. The male arrives first at the breeding colony, selects the nest site and collects nest material. After the female arrives, she forms a scrape by squatting, rotating and raking backwards with her feet. The male then brings nesting material to her.

The nests of most (but not all) eudyptids are so closely packed that no vegetation survives and the often muddy ground is bare except for nests of pebbles and bones. The centre of the colony is the most favoured site for nesting because those birds nesting on the periphery are in more danger of attack by predators than those in the centre. Owners of centre nests are subjected to pecks from the occupants of all the nests they pass. They try to stand tall and thin to keep to a minimum their intrusion into the territories they pass and thus avoid provoking attack.

As mentioned in Chapter 1, with some species a strategy of brood reduction operates when conditions are poor, with the result that the stronger chick, usually the first, survives. The eudyptids carry this strategy to the extreme, even when conditions are good. They lay two eggs with an interval of four days between them. The first is often discarded, the second is always bigger and anywhere from 20 to 70 per cent heavier. The five-week incubation period begins with the second egg and is shared for several days by both birds until the male departs after 30–35 days ashore. When he returns to relieve the female she has been ashore for up to 40 days. If both eggs are incubated, the second egg is the one most likely to hatch and this occurs up to seven days earlier than the first. On occasions when both eggs hatch, seldom do two chicks survive.

In an experiment, nests of Macaroni Penguins were manipulated to study the survival of chicks from first and second eggs. It was found that the first eggs hatched and there was little difference in survival and weight at fledging of chicks from first and second eggs. The amount of energy required by a female to produce a single egg appears to be low, but it may not be significantly greater for a two-egg clutch. A probable explanation for the continued laying of a two-egg clutch is that the extra egg represents an insurance against failure of the larger egg or first-hatched chick. Where the breeding season is short, that is in higher latitudes, the difference in egg size tends to be greater. Where the interval between eggs is increased, the difference in size between them is smaller.

Eudyptid chicks are similar in their colouring — dark brown or greyish-brown with white underparts. If frightened while still at the nest site, they hide their heads and shoulders under the belly of the male, the sole guarding parent during the brooding period. The female returns daily with food for the chick. Most eudyptid chicks join a crèche when 3–4 weeks old and they are then fed by both parents until they go off to sea at 10–11 weeks. Birds may begin to breed at the age of three years, although it is usually two or three years later than this.

Birds are separated into species by the fact that they do not breed with other than their own species. Rockhopper Penguins appear to be changing the rules. They have been found to breed with Macaroni, Royal and Erect-crested Penguins. In some cases the hybrid offspring have been found to breed also. If they continue one wonders what they will be called. Rockaronis?

# Rockhopper Penguin *Eudyptes chrysocome*

## DESCRIPTION

The specific name *chrysocome* means golden haired.

With a length of 45–58 cm, a flipper of 15–19 cm and a weight of 2.5–3.5 kg, this is the smallest of the eudyptids (similar in size to the Galapagos Penguin and larger only than the Little Penguin). The face and neck are black, the upperparts blue-black and the underparts white. A bright yellow stripe that does not meet above the bill extends back horizontally above the red eyes into long projecting fibrous yellow plumes that droop and project laterally. A black occipital crest (across the back of the crown) joins the two yellow crests which also contain black feathers. The dark orange-brown bill is short and bulbous, and the feet and legs are pink. Chicks in second down are dark greyish-brown on the head, neck and upperparts and white underneath. Fledglings are smaller than adults and the superciliary stripe is inconspicuous or absent, the throat is grey, the bill small and blackish-brown, and the eye dark brown.

## DISTRIBUTION, DISPERSAL AND POPULATION

The breeding distribution of Rockhopper Penguins is circumpolar on subantarctic and south temperate zone islands of the Indian and Atlantic oceans. They disperse northward during the non-breeding season when vagrants, usually moulting, reach the southern shores of continents in the temperate zone and New Zealand.

The population of Rockhopper Penguins breeding south of the Antarctic Polar Front was estimated in 1993 at 730 000 breeding pairs. This is 20 per cent of the world population whose total, by extrapolation, should be about 3.7 million pairs. This figure may be inaccurate because of environmental changes (see 'Threats and Conservation'). For instance, on Campbell Island south of New Zealand numbers have declined from an estimated breeding population of 1.6 million birds in 1942 to 103 000 breeding birds in 1985.

## AT SEA AND ON LAND

When going to sea Rockhopper Penguins slide and bounce down banks and cliffs, often falling over large drops without mishap. They often jump feet-first into the water, unlike other penguins that either walk or dive into it. They are gregarious at sea, keeping in contact with short barking calls, and are generally considered to feed offshore in small to medium sized groups.

They land where the sea breaks on rocks, sometimes negotiating heavy belts of kelp before scrambling ashore. They appear uncertain when they walk, but they can travel fast and sure-footedly by bouncing along with both feet held together, hence their name of Rockhopper Penguin.

## BEHAVIOUR

The Rockhopper Penguin is the fiercest and most aggressive of the penguins, not only towards other penguins but towards humans too. It attacks the legs of anyone walking through its colony and hangs on like an attacking puppy. It is extremely vocal on land, though generally quiet at night. When the penguin is approached, the black crown feathers curve up into a shining black crest and the cheeks, puffed out like whiskers, produce a cat-like appearance. Wide open red eyes and pinhead pupils show momentary curiosity.

Threats by females are less vigorous than by males. The male threatens mildly by turning its head to one side and bobbing up and down, with flippers raised for action and with short cries. In a less mild threat it jabs its open bill towards an opponent and utters harsh cries. A sitting bird may be pulled off its nest by

Fig. 35. Rockhopper Penguins at a landing surrounded by kelp.

interlocked bills. The birds grapple until one can grip the other by the nape and belabour it with its flippers. Very aggressive males may cling to their opponents as they flee and follow them through the colony, disregarding pecks from other birds.

Stones, grass and earth are collected, usually by the male, and offered to the sitting partner, who quivers and places them in the nest. Bowing may be performed by solo birds or by a pair, and often other birds join in. Standing with its bill near its feet, the bird utters a succession of deep throaty throbbing notes, the body shakes in time with the calls but the head remains still.

A bird returning to a nest hunches its shoulders, holds its body fairly upright, tilts its head and bill forward and down, holds its flippers stiffly forward and down, then pads around with a mincing gait, pivoting on its feet before settling on the nest. This action is also used by a bird relieved of nest duties and by the male immediately after copulation. Just before nest relief the partners stretch their bills towards each other and break into loud trumpeting, a series of braying vibrating yells. Neighbours often join in and direct their cries towards the newcomer. When the relieving bird reaches the nest, both birds of the pair break into vertical trumpeting with wide open bills pointed skywards, raising and lowering their flippers in time with their calls.

A single male may advertise by bowing, then swinging his head back to point skywards he shakes it ecstatically from side to side, utters pulsating cries and progressively raises his flippers. This may lead to mutual display if a female is present, in which both combine bows, vertical head-swinging, quivering and song.

Either sex solicits, always at the nest, the female by squatting, the male by beating her with his flipper. The male crowds up to his mate, nibbles her nape and flicks his flippers against her back. She subsides and lies quietly with her head raised, sometimes stretching her flippers to touch the ground. The male mounts, treads her back, drums his flippers on her flanks, moves his tail from side to side and edges backwards. He depresses his tail, the female tilts hers upwards and cloacal contact is made. The male remains still, propped in place by his flippers and with the female's bill turned into his neck. He then slides off and remains still in the shoulders-hunched attitude, after which both preen, shake their heads and relax.

BREEDING
Breeding takes place between August and March, but varies with locality, generally occurring later further south. Most colonies are

enormous and may be situated from near sea level up to 60 m above sea level and more than 1.5 km from the landing place. Colonies are often mixed with those of other seabirds such as albatrosses and shags. Sites are in rugged, rocky terrain on shores, promontories and steep slopes, but not on vertical rock faces.

Fig. 36. Rockhopper Penguin nesting site.

Rockhopper Penguins follow traditional tracks from the sea to their nests, which are sometimes sheltered from the weather by vegetation or rocks and sometimes exposed. Vegetation may be sparse or absent. Stones, tussock grass and earth are collected and carried to the nest. There may be two nests per square metre.

Older birds occupy the same nest sites year after year, but pair bonds are often broken, even when both previous partners are in

Fig. 37. Rockhopper Penguin chick begging.

the colony together. The first egg often fails during incubation and is lost from the nest during squabbles. Usually the second egg is the first to hatch, few chicks being hatched and reared from first eggs. First eggs are sometimes evicted from the nest when they are hatching. After a nest failure, there is no replacement laying.

When a chick is too large to be brooded, it stands at the nest with its parent. It joins a crèche at about three weeks, but is fed at the nest in response to parental calls, or it may chase its parents until it is fed. By ten weeks of age it has gone to sea. There is no information on the age at first breeding.

After breeding, Rockhopper Penguins moult at the nest and are ashore for three to four weeks. Non-breeders moult earlier on the edges of colonies or near landing places on the shore.

## THREATS AND CONSERVATION

Predators vary from place to place: skuas eat eggs and both they and giant-petrels take chicks and injured adults; rats eat damaged or discarded eggs or moribund hatchlings; and fur seals and hooker's sea-lions take adults at sea, especially near landing stages. Moulting Rockhopper Penguins are attacked by parasites, especially ticks. Some birds are infected by the fatal disease avian cholera, to date known only to affect penguins on Campbell Island. The drastic decline in that population appears to be associated with rising sea surface temperatures. This threat is discussed in Chapter 8.

# Fiordland Penguin *Eudyptes pachyrhynchus*

## DESCRIPTION

The specific name *pachyrhynchus* means thick bill, though this bird's bill is not as stout as that of the Snares Penguin. Where the smaller Rockhopper Penguin is the fiercest of the eudyptids, the Fiordland Penguin is the most timid.

The adult bird has a length of 55 cm, a flipper of 17–20 cm and a weight of 3.4–3.7 kg. Its head, throat and upperparts are blue-black and underparts white. A broad sulphur-yellow superciliary (eyebrow) stripe diverges from the base of the bill at the forehead and extends back over the eye to the back of the head where it develops into silky plumes that droop down the sides of the nape. Often displayed on the black cheeks are three to six white stripes roughly parallel with the superciliary stripe. There is no fleshy skin around the margin of the heavy bill, which is orange-brown, and the legs are pinkish-white. The second down of chicks is attained at about 12 days when the head, throat and upperparts change to dark brown and the underparts to white. The crests of the fledgling are short and generally lie against the head as a continuation of the yellow superciliary stripe. Eyes are dull brown to dull claret and the bill blackish-brown.

## DISTRIBUTION, DISPERSAL AND POPULATION

Fiordland Penguins are distributed in cool temperate waters around New Zealand and they breed in the South Island of New Zealand and on offshore islands. They are absent from the breeding areas from early March until June, but their distribution at that time is poorly known. Vagrants, mostly immature moulters, have been recorded on southern Australian coasts and on subantarctic islands adjacent to New Zealand.

Fiordland Penguins have a small population, estimated at 5000 to 10000 pairs, though this is difficult to verify because some breeding areas in dense vegetation are inaccessible. In 1992 it was estimated that there were fewer than 1000 nests of Fiordland Penguins per season.

### AT SEA AND ON LAND
Fiordland Penguins feed either singly or in small groups during daylight. Short barking calls keep group members in contact. It is thought that during breeding they forage inshore within 10 km of the breeding area.

### BEHAVIOUR
Displays are difficult to observe in the Fiordland Penguins' thickly vegetated forest habitat, but the vocal components of display are loud and conspicuous and similar to other eudyptids. Fighting birds interlock bills and swat each other with their flippers, while uttering harsh, low-pitched sounds. The white cheek stripes are puffed out and conspicuous during aggressive acts.

Fig. 38. Aggressive display of the Fiordland Penguin.

# Fiordland Penguin *Eudyptes pachyrhynchus*

Fig. 39. Courtship display.

## BREEDING

The breeding season begins in June with the return of the males to the breeding areas. These are in temperate rainforest along the shores of bays, fiords, headlands and — on rocky coasts — in rock falls, caves and under overhangs. Where the forest canopy is high and has a shrub layer of vines and good ground cover, there is high humidity and a wet forest floor, and temperatures are less extreme. Small colonies or solitary nests are situated anywhere from near sea level to 60 m above sea level and are rarely more than 800 m from the sea. Birds land on beaches of large boulders, and the routes to the nesting areas often follow drainage channels.

Fiordland Penguins show strong fidelity to the nest site and to partners from year to year. Nests, typically 2–3 m apart, are spread out on steep slopes of dark humus-rich soil, pockets of clay and scattered boulders. Often in the hollows at the foot of trees, nests are built of fern fronds, other plant material and stones that may be stolen from nearby nests. The area within pecking distance of the nest is the limit of territory, but breeding pairs can rarely see more than two other pairs. Because nests are well spaced, there is less aggression between males and less danger of loss of eggs.

Most clutches of eggs are completed by the end of July. Usually the second egg is the first to hatch, but chicks are hatched from a small proportion of first eggs. When this happens, the smaller chick is unable to compete for food and generally dies. The male guards the chick, which sits on his feet tucked into the brood patch, while only the female feeds the young. When the chick is too large to brood it stands or leans against its parent. At about three weeks old it wanders around the breeding grounds and eventually joins a crèche. Chicks crèche only if other chicks are near, but they return to the nest for feeding, now by both parents. Crèched chicks allopreen but fights are common.

Chicks leave for the sea in November when they are ten or more weeks old. They return to their natal colonies when three to five years old and breed first when five to six years old.

The greatest cause of nest failure is heavy rain and frequent storms. Eggs become chilled, or they are washed out of nests and chicks are drowned.

Adults take about three weeks to moult at their nest sites, while immatures moult on the edges of colonies or along rocky coasts elsewhere.

## THREATS AND CONSERVATION

The weka, a large flightless rail, is the only large natural enemy on land, taking unguarded eggs. Black flies carry parasites that infect chicks in crèches. Introduced stoats and rats take eggs, chicks and even adults; near sites of human habitation dogs are a menace and will kill adult penguins. Humans disturb nesting birds and destroy habitat by clearing or road widening, which results in rock and scree slips.

Plate 5
**Snares Penguin** *Eudyptes robustus*
1. Adult, 2. Adult, 3. Juvenile, 8. Downy young, mesoptile
**Fiordland Penguin** *Eudyptes pachyrhynchus*
4. Adult, 5. Juvenile
**Erect-crested Penguin** *Eudyptes sclateri*
6. Adult, 7. Juvenile

# Snares Penguin *Eudyptes robustus*

## DESCRIPTION

The specific name *robustus* refers to the Snares Penguin's stout bill.

With a length of 51–61 cm, a flipper of 17–20 cm and a weight of 2.8–3.4 kg, the Snares Penguin resembles the Fiordland Penguin with the head, throat and upperparts blue-black and underparts white. It is distinguished from the Fiordland Penguin in that it does not have the pale cheek stripes; the bright yellow superciliary stripe forms a laterally projecting or drooping crest behind the eye; and there is prominent pinky-white bare skin at the base of the bill. The heavy bill is orange-brown and the eye usually reddish-brown, but sometimes pinkish or even pale yellow-brown. Legs are pinkish-white. The second down of chicks, attained at 9–12 days, is medium chocolate-brown above and pale buff below. The crest of the fledgling is short and only slightly spread posteriorly.

## DISTRIBUTION, DISPERSAL AND POPULATION

Snares Penguins are birds of the temperate and subantarctic waters around their sole breeding place on Snares Island, south of

Plate 6
**Royal Penguin *Eudyptes schlegeli***
1. Adult, 2. Juvenile, 3. Downy young
**Macaroni Penguin *Eudyptes chrysolophus***
4. Adult
**Rockhopper Penguin *Eudyptes chrysocome***
5. Adult, 6. Adult, 7. Juvenile, 8. Downy young

New Zealand. Single birds, generally moulting and sometimes in company with other eudyptids, have been recorded on other islands around New Zealand and even in southern Australia. They are absent from their breeding grounds from February to August, except for moult in March/April, but their movements at sea are unknown.

The population appears to have increased since counts began in 1968 and it had stabilised in 1988 at an estimated 66 000 birds.

## AT SEA AND ON LAND

Snares Penguins forage in shallow dives near the coast of Snares Island and feed during the day in small groups. They keep in contact with short barking calls, and join mixed feeding congregations of other seabird species.

These penguins usually walk on land, but will also hop or toboggan. They land on granite points or slopes on the more sheltered eastern side of the island, following a steep route beside a watercourse to the colony. Snares Penguins actually roost in low trees.

## BEHAVIOUR

Snares Penguins have a wide range of conspicuous visual and vocal displays that are similar to those of other eudyptids, especially Fiordland Penguins. They are more social and generally interact more than Fiordland Penguins.

## BREEDING

The breeding area is occupied from late August to early February. Breeding colonies, unmixed with other penguin species, may be up to 600 m from the shore and 70 m above sea level.

There is strong fidelity to the nest site, which is defended, and pair bonds are maintained year after year. Nests are built on flat or gently sloping muddy areas, on exposed granite or among rock falls, under the shelter of the forest or scrub, and sometimes in open swampy areas. Sites are noisy and conspicuous because there

Fig. 40. Appeasement display of the Snares Penguin.

Fig. 41. Courtship display.

is little ground cover. Shade is important and colonies are eventually moved away from areas that have been flattened by the birds and their guano.

The nest is a shallow hollow rimmed with small stones, and sometimes grass and guano. The maximum density of nests is two per square metre.

As with Fiordland Penguins, the second egg is usually the first to hatch, with chicks hatched from only a small proportion of first eggs. When this happens, the smaller chick is unable to compete for food and generally dies.

The male stays with the chick throughout the guard stage and preens it regularly. The chick rests on its parent's feet tucked into the brood patch. The female returns in the early evening to feed the chick and stays till dawn.

Chicks crèche at about three weeks in crèches of up to 30 young. When they are nearly 11 weeks old they make their way to the sea in small groups, turning back frequently before diving in. The age at first breeding is thought to be six to seven years. The oldest recorded Snares Penguin was more than 21 years old.

Moult takes three to four weeks, and breeding birds moult at the nest site. Non-breeders moult at the fringes of colonies or on bare rock above the landing areas. Sometimes they moult away from Snares Island in New Zealand and southern Australia.

## THREATS AND CONSERVATION

Adult Snares Penguins are occasionally taken by hooker's sea-lions and leopard seals. Fledglings are killed by giant-petrels near the shore or at the launching rocks. Skuas are not known to take chicks, but middens used by skuas contain egg shells and chick carcasses. No harvesting of penguins is known and human activity is probably not a menace.

# Erect-crested Penguin *Eudyptes sclateri*

## DESCRIPTION

The specific name *sclateri* refers to P. L. Sclater (1829–1913), Fellow of the Royal Society and secretary of the Zoological Society (London), in whose honour the Erect-crested Penguin was named. The common name indicates its distinctive feature: other eudyptids all have drooping crests.

With a length of 67 cm, a flipper of 19–23 cm and a weight of 4–5 kg, the Erect-crested Penguin is larger than the Snares and Fiordland Penguins which it resembles with its black upperparts, head and neck, and white underparts. The broad pale golden-yellow superciliary stripe rises from near the bluish-white bare skin at the base of the bill and continues obliquely over the eye to form a long brush-like erect crest of silky feathers. The eyes are brown, the feet and legs pink and the long slim bill is brownish-orange. Chicks in second down have dark brown upperparts, head and neck, and white underparts. The crest of the fledgling is smaller than that of the adult and the bill is dull black with a light tip.

## DISTRIBUTION, DISPERSAL AND POPULATION

Erect-crested Penguins live in subantarctic and cool temperate waters around New Zealand. They breed on islands around southern New Zealand (Bounty, Antipodes and Auckland islands), as well as on small neighbouring islands with little vegetation or soil. They are absent from their breeding places from May to early September, but their movements at sea are unknown. Vagrants, some of them immature moulters, reach the coasts of New

Zealand and southern Australia and other subantarctic islands, including one recorded on the Falkland Islands in the South American zone.

Their population is estimated at about 200 000 breeding pairs.

## AT SEA AND ON LAND
Breeding Erect-crested Penguins forage close to their colonies and keep in contact with short barking calls. Groups of up to 300 birds have been observed close offshore. On land they usually walk or hop.

## BEHAVIOUR
Erect-crested Penguins are gregarious during breeding and moult. Their social repertoire is similar to that of other eudyptids, especially Fiordland and Snares Penguins, but they are generally less aggressive. They are unique among eudyptids in being able to raise and lower their crests, but the function or effect of this is not known.

## BREEDING
Erect-crested Penguins return in early September to breed in large colonies. These are established in rocky terrain — on beaches strewn with boulders, rocky flats, slopes and ledges — from just above sea level to 70 m above sea level. Sometimes colonies are mixed with Rockhopper Penguins or shy albatrosses. There appears to be fidelity to the nest site and to the partner. Territories extend to within pecking distance of the nests, which are at a density of one per 1.4 square metres. The nest is a shallow hollow, rimmed with small stones, and sometimes grass and guano, built on open, level or gently sloping ground. Eggs are laid in October. The one egg incubated, almost always the second-laid, is partly held on the parent's in-turned toes, with the adult sitting at a 45° angle to the nest or fully prostrate over it. Chicks stand on their parents' feet tucked into the brood patch under the belly. They crèche like other eudyptids, and fledglings leave by mid-February.

There is no information as to the age of first breeding.

Adults moult at the nest site and depart in mid-April, while immatures moult at colonies or along rocky coasts elsewhere.

## THREATS AND CONSERVATION
The breeding habitat of Erect-crested Penguins is essentially unchanged and under no threat. Adults vigorously defend their nests and chicks from skuas. Northern giant-petrels that gather at the time of fledging eat dead birds, but have not been seen to kill.

# Macaroni Penguin *Eudyptes chrysolophus*

## DESCRIPTION

The specific name *chrysolophus* refers to the gold crest. Macaroni Penguins, with their long, drooping and spikey crests of orange, yellow and black plumes rising from the centre of their foreheads, were named after a bizarre hair fashion known as the 'macaroni'.

With a length of 71 cm, a flipper of 20 cm and a weight of 5–6 kg, the Macaroni Penguin is the largest of the eudyptids (except for the very similar Royal Penguin). The head, cheeks, throat and upperparts are blue-black and the underparts white. There is an angled line of separation between the black throat and the white breast. This differs from other eudyptids that have a straight line. A line of black feathers separates a central orange-yellow patch on the forehead from the bill. Conspicuous long chrome-yellow, orange and black plumes rise from the central patch and project backwards along the crown or droop behind the eye. The massive bill is red-brown and the bare skin at its base forms a bright pink triangle. The eye is red and the legs bright pink. Chicks in first down are dark grey-brown and white underneath. The head plumes of the fledgling are composed of scattered small yellow feathers on the forehead. The fledgling's eye is dull brown and the bill black-brown.

Adult Macaroni Penguins can be distinguished from Royal Penguins, which have white or grey cheeks and throat, although in some breeding places there is a mixture of the two throat colours. Macaroni and Royal Penguins are the only eudyptids with crests that meet on the forehead, giving them a somewhat rakish look, the antithesis of the stately Emperor and King Penguins.

## DISTRIBUTION, DISPERSAL AND POPULATION

Macaroni Penguins live in subantarctic and Antarctic waters north of the pack-ice. Their breeding colonies, the most southerly of the eudyptids, are on the Antarctic Peninsula and on Antarctic and subantarctic islands.

Macaroni Penguins are migratory outside the breeding season, but their movements at sea are not well known. Rare vagrants have been recorded in South Africa and Australia.

The population of Macaroni Penguins breeding south of the Antarctic Polar Front (99 per cent of the total population) was estimated in 1993 at 11.72 million breeding pairs.

## AT SEA AND ON LAND

The diving patterns of breeding female Macaroni Penguins were followed with the use of time/depth recorders. While males brooded the chicks (one per breeding pair), females left soon after dawn and spent more than 12 hours each day foraging at sea. Dives ranged between 30 m and 70 m, but some reached 100 m. The period under water rarely exceeded two minutes. Towards the end of the brooding period, females needed to consume 1500 g daily to sustain themselves and their chicks. Sometimes they stayed at sea for one or two nights, and dives then were short and shallow. Whatever the length of time spent at sea, their return to shore was always in daylight.

Macaroni Penguins land on rocky beaches exposed to heavy breakers, and walk or hop to their colonies. Where the ground is level, movement of the birds causes erosion and the rain forms muddy depressions. Any vegetation on the way is destroyed by the passage of the birds.

## BEHAVIOUR

Macaroni Penguins loafing on beaches stay closely packed. Their behaviour is similar to that of other eudyptids.

BREEDING

In September or October, the date varying with the distance southward, birds return to their breeding islands. Colonies with hundreds of thousands of birds together are located up to 300 m above sea level. They are situated on steep rough ground, lava flows, scree slopes and rock falls, among hills, and in caves and amphitheatres. To reach the higher sites, birds have to climb up steep rocky slopes or up a stream, or they may have to traverse hundreds of metres across steep scree slopes.

Site and mate fidelity are strong. Nests lined with pebbles and rock fragments are pecking distance apart (30–40 cm). Egg laying is highly synchronised. Although the adult lies prone over the first-laid egg, seldom if ever is this retained; most are lost either before or at the laying of the second egg. The peak of aggressive action between neighbouring pairs occurs when territories are being established. As this largely abates with egg-laying, it is not the sole cause of egg rejection. After the second egg is laid, the colony changes from noisy confusion to strange quietness as the males depart and the females begin incubation.

Breeding success is lower in those colonies on flat ground, which are susceptible to flooding, and those on slopes or below cliffs, which are exposed to heavy rain, snow-melt or landslides.

Breeding adults moult at the nest site, non-breeders at their natal colony. Moult takes about four weeks.

THREATS AND CONSERVATION

Skuas, sheathbills and kelp gulls take eggs, and giant-petrels take isolated fledglings or those going down to the sea.

# Royal Penguin *Eudyptes schlegeli*

## DESCRIPTION

The specific name *schlegeli* was given in honour of Professor H. Schlegel (1804–1888), Director of the Leiden Museum.

Royal Penguins are often considered to be a sub-species of Macaroni Penguins and differ from them only in the amount of black feathering on the cheeks and throat. In Royal Penguins the face is white or pale grey from the crest to the throat. Measurements for Royal and Macaroni Penguins overlap, those of Royal Penguins being given as a length of 65–75 cm, a flipper of 22.5 cm and a weight of 5–6 kg. This makes Royal Penguins slightly larger than Macaroni Penguins, but this variation may be the result of a larger sample of measurements. The chicks of both species in second down are indistinguishable. Fledglings are distinguished by the lighter face of the Royal Penguins. On Macquarie Island some breeding birds have black faces and others are intermediate between black and white.

## DISTRIBUTION, DISPERSAL AND POPULATION

Royal Penguins breed only on Macquarie Island and adjacent islets. They are migratory, the last birds leaving by May. Their precise wintering areas are unknown, but are assumed to be in subantarctic waters. Vagrants have been recorded in Australia (mainly Tasmania), New Zealand, Heard Island and subantarctic islands in the Indian Ocean.

The population of Royal Penguins in 1984–85 was estimated at 850 000 pairs in 57 colonies.

## AT SEA AND ON LAND

The food and the manner in which it is taken is similar to that of the Macaroni Penguin. Royal Penguins drink from freshwater streams during the breeding season and use creeks as access routes to inland colonies.

Fig. 42. Royal Penguins using a creek for access.

## BEHAVIOUR

Behaviour and displays are similar to those of other eudyptids.

## BREEDING

Royal Penguins arrive in mid- to late-September at colonies situated along the shore, on scree slopes or among hills up to 1.6 km inland and 150 m above sea level. Fidelity to site and mate is strong. Colonies can be huge, with nests tightly packed at densities of 2.4 per square metre. The breeding routine is similar to that of Macaroni Penguins.

Royal Penguin females exhibit deliberate infanticidal behaviour by scraping the first egg away with their feet or by removing it with their bills, usually on the day *before* the second egg is laid.

Some precocious young may breed at five years of age, but adolescence is prolonged and some do not breed until 11 years old.

Fig. 43. Royal Penguins with chick.

Fig. 44. Pair bonding.

## THREATS AND CONSERVATION

Skuas and wekas, the latter introduced to Macquarie Island, prey on Royal Penguin eggs and chicks, and giant-petrels may take full-grown birds. Elephant seals that invade colonies near the shore destroy adults and young.

## Chapter 5

# Genus *Megadyptes*

## Yellow-eyed Penguin *Megadyptes antipodes*

### DESCRIPTION

*Megadyptes* means big diver and *antipodes* was used by northern hemisphere biologists to refer to islands on the opposite side of the world to Europe. Yellow-eyed Penguins are not recorded on the islands known as Antipodes. The Yellow-eyed Penguin is the only one in this genus.

This medium-sized penguin is 66–78 cm in length, its flipper 20–22 cm, and its weight 4.5–6 kg. The male is about 500 g heavier than the female before she lays; in other species the female is the heavier at this time. It is a big penguin to occur so far north, though similar in length to the Gentoo Penguin. When standing it is the third tallest penguin (after the Emperor and the King Penguins) and fourth heaviest (after the Gentoo Penguin). In the adult the feathers of the crown, sides of the face and chin are pale yellow, each feather with a central black streak. Instead of a crest, a broad band of yellow feathers extends from the base of the bill, around the eyes and encircling the head high up on the nape. The remaining upperparts are slate blue and the underparts white. As its name implies, the Yellow-eyed Penguin has straw-yellow eyes. The bill is long and slender, red-brown and pale cream, and the feet are pink. Chicks in second down are uniformly light brown. Fledglings do not have the yellow band of feathers around the hindcrown, the chin is white and the eyes pale grey-yellow.

DISTRIBUTION, DISPERSAL AND POPULATION
The Yellow-eyed Penguin lives on the south-eastern coasts of the
South Island of New Zealand, and on Stewart, Campbell and
Auckland islands. It does not share its breeding grounds with
other penguins. It does not migrate and is present at its breeding
areas all year. Fledglings, and sometimes adults, disperse northwards up to 600 km, but return to their natal areas during their
first year.

One penguin, released 87 km from its breeding area, returned
in 48 hours. Another, released 350 km away, returned in 17 days,
showing remarkable powers of navigation and, at 20 km per day,
sustained effort.

The Yellow-eyed Penguin has the smallest population of all
penguins.The total number including non-breeding birds, was
estimated in 1993 to be 3666–4500 individuals which included
1100–1350 breeding pairs. It is endangered on the New Zealand
mainland and rare, possibly threatened, throughout the rest of its
range.


AT SEA AND ON LAND
During the day birds mainly feed individually, occasionally staying
at sea overnight. When hooker's sea-lions are present, Yellow-
eyed Penguins form small groups and then make a dash for the
beach, porpoising fast. Landing on rocky shores and platforms or
on sandy beaches, they walk or struggle up the slipping sand of
steep sandhills to their nests and claw their way up to ledges on
the cliffs.


Plate 7
**Little Penguin** *Eudyptula minor*
1. Adult (Australia), fresh plumage
2. Adult (Australia) worn plumage,  3. Juvenile,
4. Adult (New Zealand) fresh plumage,  5. Adult (White-flippered) fresh
plumage, 6. Downy young
**Magellanic Penguin** *Spheniscus magellanicus*
7. Adult male

## BEHAVIOUR

In winter or when they are not breeding, Yellow-eyed Penguins are gregarious, congregating on the upper beaches of major breeding areas before and after foraging at sea.

They are extremely wary of humans and rarely land on a beach if humans are in sight, so hides are essential for observation of displays.

In Aggressive Behaviour, birds threaten each other with hunched shoulders, stares and glares, gaping and open yell (Figure 45) and they utter harsh calls. If this threat is insufficient to deter an intruder, the defender may charge, pecking at its opponent, but usually falling well short of making contact. The birds stand facing each other, bills almost touching, before they peck each other and strike with their flippers.

Fig. 45. Gaping and Open-yell.

Plate 8
**Magellanic Penguin *Spheniscus magellanicus***
1. Juvenile, 2. Downy young
**Humboldt Penguin *Spheniscus humboldti***
3. Adult, 4. Juvenile
**Galapagos Penguin *Spheniscus mendiculus***
5. Adult, 6. Juvenile
**African Penguin *Spheniscus demersus***
7. Adult, 8. Juvenile

To avoid conflict and to appease a possible opponent, birds assume a non-aggressive stance by standing tall, just as other penguins do when passing close to occupied nests. A bird (usually the male) walks with hunched shoulders rapidly past another bird, then salutes or bows by holding his bill vertically and his neck fully extended and flippers forward for about five seconds. He follows this by lowering his neck and bill and looking over his shoulder. He may give a sheepish look in which the body and flippers are held similarly to the Salute but the bill is dropped parallel to the neck. The Gawky Attitude is given in response to the Salute (Figure 46).

Fig. 46. Salute (left) and Gawky Attitude (right).

Sexual behaviour is displayed with throbbing, shaking, half-trumpeting (Figure 47) and the Ecstatic Display (Figure 48), accompanied by loud calls, which are much more musical than the harsh cries of most other penguins. Most calling occurs during the breeding season, contact calls being uttered at other times, but breeding areas are quieter than those where birds breed colonially.

Allopreening takes place between mates and their chicks.

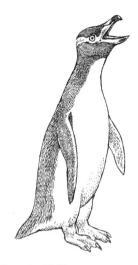

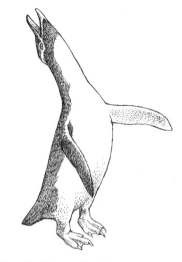

Fig. 47.  Half-trumpeting.     Fig. 48.  Ecstatic Display.

BREEDING

Birds may prepare for breeding as early as May, but eggs are not laid until September or October. Historically breeding areas were in coastal forests, but most of these have been cleared. Yellow-eyed Penguins are only loosely colonial, the least so of all penguins. These days, nests are sited on sea-facing slopes, in the remains of forests, in gullies, on hills and on cliff tops. They range from near sea level to 250 m above sea level and can be up to 1 km inland. They are never in caves or burrows. Nests need shelter from direct sunlight to protect the birds from heat stress. They are well hidden in dense vegetation and scattered over a wide area, but most are within earshot of each other. Where birds attempt to nest in sight of each other, one or both pairs will always fail.

Fidelity to the general nest areas (though not to the nest site) is strong and a territory of about 1.5 ha is defended during breeding and sometimes throughout the year. There is little interaction between neighbouring pairs. Density is one nest to 1.5 ha in forest, but may be higher in more dense vegetation.

The pair bond is long lasting once a pair has settled down to breeding for several years, but divorces do occur. Both birds build the nest which is a shallow bowl on the surface with a substantial lining of twigs, tussock grass, leaves and plant stems. Usually two or three nests are built. The two eggs laid in September/October are similar in size. The pair shares in the incubation of both eggs for 39–51 days. This is the most variable incubation period of all penguin species. It is the result of some individuals failing to cover the clutch completely for more than a week from the laying of the second egg, the time at which incubation normally begins. This delay in beginning incubation appears to result from food shortages. When food is plentiful the incubation period is shorter.

Though there is an interval of four days between eggs, they hatch on the same day and there is no competition between the siblings. Chicks are guarded continuously by the parents for 40–50 days and siblings are fed equally. There is no strategy to reduce the brood size when food is short (such as feeding the most aggressive chick). Instead, both chicks fail equally, although the early death of one would contribute to the survival of the other. Should one parent die, a single parent is capable of rearing a single chick. Chicks do not form crèches, but they may move to a more sheltered site when four or five weeks old. They do not fledge until 14 weeks old in February/March, later than most other penguins. Because the time taken for chick-rearing is so long, there is no second clutch.

Some birds may breed at two years of age, but most are three or more. Some birds are known to have survived for 20 years.

Breeding success varies from year to year. There is often more than one chick per pair, which is a high breeding rate for penguins, but 70–75 per cent of these young do not survive their first year at sea.

Breeding birds moult in their nesting area between February and April and are ashore for three to four weeks. Immature birds tend to breed in the area where they have moulted.

## THREATS AND CONSERVATION

The Yellow-eyed Penguin has suffered a serious decline through the loss of its original habitat, the coastal forests, by logging; by cattle and sheep trampling nests; and by predation by introduced cats, dogs, pigs, stoats and ferrets. Predation is high where nesting areas are close to habitation and the land is cleared for farming. Even though the Yellow-eyed Penguin is very wary of humans, this has not saved it.

The island populations are not subjected to the same stresses as those on the mainland, but it is believed that migration from the islands to replenish the declining mainland populations does not occur. The mainland populations are estimated to have declined by 75 per cent in the last 40 years.

At sea, Yellow-eyed Penguins have to contend with barracouta, their major marine predator which damage legs and feet, as well as with sea-lions and sharks. In the Otago region they have been drowned in set fishing nets and efforts are being made to have set nets banned.

Breeding adults will take readily to large nest boxes placed in suitable areas. Large predator-free reserves of suitable habitat are needed if the Yellow-eyed Penguin is to survive on mainland New Zealand.

## Chapter 6

# Genus *Eudyptula*

## Little Penguin or Blue Penguin
### *Eudyptula minor*

DESCRIPTION
*Eudyptula minor* (good little diver) is the only species in this genus.

At 40–45 cm long, with a flipper of 11–14 cm and a weight of about 1 kg, the Little Penguin is the smallest of all penguins. The name Little Penguin has displaced the previous common name of Fairy Penguin. Anyone who has had close contact with the species finds its aggressive behaviour not at all fairy-like. The Little Penguin is known in New Zealand as the Blue Penguin. The White-flippered Penguin, also of New Zealand, is regarded by most as a sub-species, the greatest difference being that it has white on both edges of the flippers whereas the Little Penguin has white only on the trailing edge. All are blue-black to blue-grey above and white underneath. Unlike all other penguins, the Little Penguin does not have any colours or crests on its head. The iris is silver-grey, bluish-grey or hazel, the bill black and the feet pinkish-white. Chicks in second down are dark grey-brown above and white underneath, with the dark grey iris changing to pale grey. Fledglings resemble adults except that they are brighter blue and have shorter bills.

On Phillip Island in south-eastern Australia, the Little Penguin has been the subject of a continuing study since 1967, the longest continuous year-round study of a single penguin population. It is the penguin most commonly seen by tourists, the 'Penguin Parade' on Phillip Island having achieved world status as a tourist attraction. Half a million visitors annually see the birds emerging from the sea at dusk under the full glare of lights, to which they have become accustomed.

## DISTRIBUTION, DISPERSAL AND POPULATION

The Little Penguin lives in southern Australia and New Zealand in seas that are warm compared with those of the cold Antarctic. In New Zealand it breeds all around the mainland and on offshore islands and the Chatham Islands. In Australia it breeds mostly on offshore islands, where it is less at risk from land predators including humans, although colonies are also situated at isolated and protected sites on the mainland, such as at the foot of cliffs, on breakwaters and other man-made structures. The long-established population on Penguin Island in Western Australia is isolated from other penguin colonies to the south and east, and the birds are larger.

Adults are regarded as sedentary and throughout most of their range Little Penguins are present at their colonies all year round, although numbers decrease following moult and in winter. Journeys of more than 1000 km from the natal area have been recorded for juveniles immediately after fledging. From Phillip Island the movement is predominantly westward, perhaps because food is more abundant there.

Using tiny radio transmitters fixed to the penguin's back together with shore-based and aircraft receivers, it has been shown that Little Penguins in Victoria travel up to 700 km from the burrow and stay within 20 km of the coast. On day trips they remain within 15 km of the burrow and 9 km of the coast. One bird travelled at least 113 km in 34.5 hours, an average speed of nearly 3.3 km/h. During winter adult birds forage more in Port Phillip Bay than in the open sea.

Many colonies contain thousands of pairs and the total

Australian population has been estimated at at least several hundred thousand birds. Little Penguins are in no immediate danger, although some colonies have dwindled.

## AT SEA AND ON LAND

The Little Penguin forages during the day on small shoaling fish under 12 cm long and takes the most abundant. Small fish are caught and swallowed head-first under water, but those more than 26 mm long are brought to the surface to be swallowed. The movement of the fish and its flash of light colour provide the stimulus for the feeding chase. Samples of food have been obtained by stomach flushing with water in conjunction with a study of mid-water trawling.

Fig. 49. Little Penguins heading to sea before dawn.

Fig. 50. Little Penguins at Penguin Island, Victoria.

The Little Penguin sometimes porpoises when swimming, but more often it comes to the surface for a moment, breathes and dives again. At sea, birds lie low in the water often in small groups.

The Little Penguin is the only penguin that is wholly nocturnal on land, departing and returning under cover of darkness, although on isolated islands and in rockfalls on the cliffs bordering the Great Australian Bight it will come ashore in daylight. Large rafts form offshore until, at dusk and for up to two hours afterwards, birds come in to land. They appear quite suddenly from out of the surf to stand on the shore before walking across the beach to follow well-worn paths to their burrows. They will toboggan on the sand if stressed. Where shores are rocky, they jump from the pounding waves onto rock platforms and scramble upwards out of reach of the waves. Even granitic rocks have become scored by the toenail scratches inflicted by penguins over countless years. Little Penguins walk and hop down steep slopes but remain upright. During the day they remain inside their burrows, but they will walk about inside caves.

## BEHAVIOUR

Displays are noisy and contact calls appear to promote social cohesion both on land and at sea. At night, especially after the birds have returned from sea during the breeding season, the noisy drumming of the male's flippers on the female's back indicates copulation. All other penguins point their heads vertically as part of their Mutual Display. In a New Zealand study the Little Penguin did so only during the Advertising Display. In Australia, where the birds have been watched for many years displaying outside their burrows, this vertical pointing has never been observed; the restricted space inside the burrow would preclude such a display. Apart from this possible geographical difference, the displaying Little Penguin in both areas raises its head and points its bill upwards above the horizontal but below the vertical.

In threat behaviour the Little Penguin uses a variety of stances. With its head and body held low and its flippers held upwards away from the body, it may point its bill towards its opponent. With upright body, the feathers of its head erect and flippers outstretched, the bill and white underparts are directed towards its opponent while it growls or brays. Noisy threats are more effective than silent ones.

Aggressive interactions usually begin with behaviour that poses the lowest risk to the bird displaying. If the opponent is deterred by the threat, it will either hunch its shoulders and walk away, or remain but look away. If the opponent is not deterred, aggressive display may intensify and both birds are then at risk. In males it may lead to fighting, which can result in flesh wounds and sometimes even in the loss of an eye.

Pairs allopreen, but even a group of birds roosting together and resting with closed eyes may preen one another if the bill of one touches the body of another.

Although Little Penguins are active during the night, they also sleep for short periods, as well as sleeping during the day in their burrows. Periods of sleep are short, but the total amount is equivalent to that of other birds. They sleep either lying prone or standing, with indrawn head and slightly upward pointing bill.

## BREEDING

The start of the breeding season is variable and differs between colonies, even those at the same latitude. It begins earlier in Western Australia (April) where the climate is warmer than in the eastern states and New Zealand, where it usually begins in late winter. Little Penguins occasionally breed as solitary pairs, but generally they form loose colonies, situated up to 50 m above sea level and 300 m inland.

Site fidelity is strong in Little Penguins and burrows are visited throughout the year. Mate fidelity is also strong, though divorce may occur within a pair that has bred successfully in previous seasons.

Both members of a pair dig the burrow using their bills and feet. Alternatively, nests may be built under rocks, bushes or buildings, or in caves; anywhere that provides shelter. Nests in caves are built 1–2 m apart but the distance between burrows may be slightly more than 2 m. Grass, plant material and seaweed line the nest at the end of the burrow.

In Australia males are usually first at the nest site, but in New Zealand an unmated male will pair with a female before obtaining a nest site. Sometimes shearwaters and fairy prions nest in penguin colonies, but the penguin is the dominant one and will oust a shearwater that has taken over its burrow. Where a shearwater chick is unfortunate enough to have a penguin oust its parent, it will be brooded by the penguin but not fed. The chick is unable to give the right stimulus for the penguin to respond and it does not survive.

Fig. 51.  Little Penguin nesting in a cave.

A small area near the mouth of the burrow is defended. Although the Little Penguin is small it is quite savage and capable of protecting itself, partly by bluff and partly by physical contact — biting and battering with its flippers.

Egg-laying is not synchronous and takes place over a prolonged period. Some birds replace lost clutches and, when breeding has begun early in the year, others continue to a second clutch after a successful first. Such second clutches can also be reared successfully. Both parents alternate to incubate the two eggs for five weeks, then brood and guard the chicks for a further two to three weeks, after which both go to sea daily.

Bigger chicks come out of their burrows at dusk to meet their returning parents. They will beg greedily and noisily, almost knocking over the adult birds until they are fed near the nest. When they continue to clamour for food, the parents may move away and shelter in another burrow. Although crèching is not usual, sometimes a group of neighbouring chicks will shelter in the same burrow, but birds nesting in caves are much more likely to form small crèches.

Fig. 52. Little Penguin about to fledge. (Note band on flipper.)

Fledglings leave for the sea when they are eight or more weeks old and disperse widely during their first year or two of life. Occasionally young have been recorded moulting at a great distance from the natal area, but of the hundreds of banded birds later found breeding, only two did so away from their natal colony. The Little Penguin can breed at the age of two, but more often at three, and one bird was recorded still breeding at the age of 21.

Breeding success appears to be correlated with sea surface temperatures in Bass Strait: cooler temperatures associated with west winds result in earlier breeding and higher breeding success. It is suggested that the west winds bring cooler water with higher nutrient concentrations into Bass Strait, aiding the growth of food. The strength of the west winds in Melbourne in July may provide the best indication as to how well the penguins on Phillip Island will fare during the coming breeding season.

Unlike at Phillip Island, breeding success in Western Australia does not differ with changes in the surrounding waters (brought about by the Leeuwin Current), even though adult weights are low when the current is warm and weak. The reverse applies when the current is stronger.

When there is little food, breeding everywhere is delayed and may be unsuccessful. Over the period 1986–88, breeding was less successful when less food was brought ashore. At the same time fewer fish were caught in mid-water trawls off Phillip Island. When there are fewer fish, Little Penguin diets are high in squid, which has a lower food value for chicks. When eggs do not hatch synchronously, the younger chick is at a disadvantage and may not survive a food shortage.

Mortality per year is estimated to be 25 per cent in adults, while 69 per cent of fledglings do not survive to maturity. It is also estimated that 20.2 per cent of males and 17.6 per cent of females are responsible for all of the following generation.

After breeding, adults spend anything from a few weeks to several months at sea to fatten up for the moult. They may double their body weight and become so fat that it is difficult for them to make the climb up sandy slopes or steep paths to their usual burrows, and they may seek shelter under rocks or bushes or whatever other protection is available. Moult takes two to three weeks.

THREATS AND CONSERVATION
Large-scale oceanic events, such as ENSO events, probably affect food production. If long-term oceanographic changes are taking place in Bass Strait, it is likely that there will be corresponding changes in fish populations, but there is no way to combat this threat. One human threat could be combated by refusing licences for the commercial harvesting of fish which are also suitable for penguins. Such harvesting has been expanded in response to increased demand for human consumption, for bait and for pet foods.

Most mortality occurs in birds weakened by lack of food, combined, in the younger bird, with internal and external parasites. To breed successfully Little Penguins must anticipate the capacity of local food resources for months ahead. They are not always accurate and consequently they have a breeding failure.

The effect of natural predation by sharks, gulls, sea-eagles and New Zealand fur seals has not been studied, but on Penguin Island in Western Australia king's skinks prey on penguin eggs and may lower breeding success.

Problems caused by humans are not easy to overcome. Penguins fall prey to foxes, dogs and possibly cats in some places, and to stoats in New Zealand. Even colonies facing the ocean at the foot of high cliffs are not safe from the introduced predators that can find their way down. Birds become entangled in waste plastic and fishing lines, and oil spills cause further deaths. Little Penguins are killed on the roads; burnt in grass fires; accidentally crushed in burrows by people, vehicles and farm animals; and illegally poached for food or crayfish bait.

When the bodies of dead Little Penguins from eastern Australia were examined for heavy metals and pesticides, they showed concentrations below the level likely to cause harmful physiological effects. Nevertheless, those specimens found close to cities showed greater contamination and indicate that greater control of contaminants is necessary. Other penguin species examined in the higher (colder) latitudes, showed only low levels of contamination.

The greatest threat in the past was the destruction of habitat as it was converted into farms and residential developments, thus reducing the area available for penguins. In many places this encroachment into penguin habitat has now been halted, and in some places reversed. A housing development near the remaining colony on Phillip Island was stopped in the early 1980s, the land resumed, 180 houses removed and the habitat restored, including the provision of artificial nest sites. Little Penguins take readily to nest boxes or pipes sunk into the ground.

Foxes are the greatest continuing threat, despite a program that attempts to eliminate them. Road kills have been greatly reduced by restricting road use around the nest sites to essential traffic at night, but there is evidence of occasional bursts of vandalism that leave dead penguins strewn on the roads.

On southern Australian coasts periodic wrecks of penguins, usually less than six months old, are possibly the result of poor food availability and stormy weather. These wrecks may cause concern to the general public, horrified to see dead bodies on the beaches, but such wrecks are not frequent and are probably part of the normal hazards of a penguin's life.

## Chapter 7

# Genus *Spheniscus*

## INTRODUCTION

The generic name *Spheniscus* means wedge-like and refers to the penguin shape.

This so-called warm weather group, the spheniscids, contains four species: Magellanic, Humboldt, African and Galapagos Penguins. All are black and white, with either one or two black bands across the chest, and are medium sized apart from the smaller Galapagos Penguin. They do not have crests or colours, but their bare facial skin is often pink from the blood under the surface and it helps the birds to lose heat. The behaviour of all four species appears to be similar, and it is given in detail for the most studied, the African Penguin.

Although all four species resemble each other, they have only been known to interbreed in zoos. Galapagos Penguins are well separated from African Penguins by the South Pacific Ocean and are found 5° of latitude north of Humboldt Penguins. Humboldt and Magellanic Penguins overlap in breeding range for about 300 km on the coasts of Peru and Chile, but even when they share the same small islands, they are usually found in separate areas. There are occasional reports of interbreeding in the wild and they indicate that hybrids appear to have lower survival and breeding success. There can be no human confusion in identification because Magellanic Penguins are the only ones that have two complete black bands across their chests.

All adult spheniscids have a clear pattern of spots on the belly, which can be used for individual identification when small numbers are studied as the markings do not change between moults. The bare black and pink pattern of the facial area forms a mask and may differ from one side of the face to the other. After moult the bare skin is covered by small white feathers but, except for Magellanic Penguins, these are soon lost.

Spheniscids swim low in the water with the head and part of the back exposed and the tail often cocked. Large groups sometimes bathe near the shore before setting out on foraging trips. If they stay overnight at sea, they sunbathe on the water in the morning.

They forage during the day, leaving at dawn and returning before nightfall, although some may remain at sea for several days. Most often they forage individually, but sometimes they hunt together in small groups of fewer than 20 birds. They herd schools of fish together, then seize and swallow, continuing until they have to rise to the surface to breathe. When they feed in groups with other bird species there is often benefit from the different foraging strategies which tend to cause the fish to school (cormorants foraging from the side and boobies diving from above).

Apart from Magellanic Penguins, spheniscids may breed at any time of the year depending on food abundance, laying a second clutch even when the first has been successful, and sometimes even a third. If food is scarce, as in ENSO events, they may not breed at all.

All four spheniscids dig burrows in guano or earth or nest under vegetation or in rock crevices, wherever there is shelter and shade. There is strong fidelity to site and mate. Nest building, incubation, brooding and guard duties are shared by both parents. Egg-laying is synchronised within colonies but not between one colony and another. The two eggs are similar in size with an interval of three or four days between laying. In very hot weather spheniscids pant deeply and they may stand to shade the eggs rather than lie over them. Incubation takes 38–42 days and may begin soon after the first egg is laid, but not in earnest until the second one, with an interval of two to three days between hatching. Chicks do not run after parents to be fed. They fledge as early as eight weeks old, but usually later. If there is little food they may take up to 17 weeks.

Allopreening takes place with all spheniscids. Galapagos Penguins undergo prenuptial moult, but the other three species follow the usual pattern of moult after breeding.

# Magellanic Penguin *Spheniscus magellanicus*

## DESCRIPTION

The specific name *magellanicus* honours Ferdinand Magellan whose expedition was credited with the first reports of penguins.

The adult Magellanic Penguin has a length of 71 cm, a flipper of 18–20 cm and a weight of about 4 kg, and is the only spheniscid with two complete black bands across its chest. The black cheeks, throat and narrow cap are divided by a broad white superciliary stripe from the black upperparts. A broad black band crosses the chest to join the black shoulders. Below this a narrow white band extends from the chest down the flanks. A second black band across the chest also extends down the sides, thickens in front of the flippers and continues downwards to end as blackish blotches at the top of the upperlegs. The rest of the underparts are white, sparsely blotched with black. The stubby bill is black, the iris brown and the pink legs and feet are blotched with black. Bare pink skin extends to the eye and from the bill during the breeding season. Chicks in second down are brown above and white underneath. Fledglings are similar to adults but the plumage is greyer.

## DISTRIBUTION, DISPERSAL AND POPULATION

Magellanic Penguins are less warm-weather penguins than the other three species. Colonies are located from the Patagonian coast of Argentina, around the southern tip of South America to the coast of southern Chile. In the Falkland Islands they breed in mixed colonies with Rockhopper and Gentoo Penguins.

Magellanic Penguins migrate from the most southerly colonies between April and August, but are otherwise sedentary. The breeding distribution has expanded northwards, perhaps due to changes in the location and abundance of food.

One bird identified by its band was recovered 3300 km from its place of banding, perhaps the longest distance travelled by a non-aerial bird. Rare stragglers that reach Australia and New Zealand are usually moulters.

The world population in 1975 was estimated to be between one and two million birds.

## AT SEA AND ON LAND

Magellanic Penguins fish in groups and have been seen fishing several hundred kilometres offshore. In the Falkland Islands there is a great variation in the prey taken by Magellanic Penguins from different colonies as close as 60 km apart. Magellanic Penguins porpoise occasionally and dive through the waves. They call when floating and are especially noisy on land, particularly at night. They walk on land and moulting birds may skitter along on all fours when alarmed.

## BEHAVIOUR

Magellanic Penguins are aggressive and capable of inflicting deep wounds with their formidable bite. Though shy on land, in the water they are fearless of humans.

## BREEDING

The breeding season of the Magellanic Penguin begins in early September or October, the further south the later. Colonies are situated up to 70 m above sea level on mainland coasts and islands that are grassy or forested, on tiers of cliff faces, or on banks and sandhills. Nests, lined with bones, twigs and litter, are built in burrows dug in sand or shingle, or they may be shallow scrapes under bushes or occasionally among rocks. Magellanic Penguins do not normally re-lay after failure of the clutch.

Burrows are deep, incubation takes about 40 days, and chicks remain in the burrows until they fledge. They do not crèche and they fledge when aged between 9 and 17 weeks, depending on food provisioning.

Parents distinguish between the first and second chicks and there is higher mortality of the second. If chicks miss one feed, it can mean death as sometimes they must fast for several days between feeds. Breeding adults moult in the breeding area, but juveniles travel to where food is plentiful and moult there, some-times forming a dense band along the shore.

## THREATS AND CONSERVATION

Most threats to Magellanic Penguins are the result of natural causes, starvation being the major cause. El Niño events bring food shortages and torrential downpours. Burrows are flooded and soaked chicks die of hypothermia. Burrows can be collapsed by rain, sheep or guanacos (a type of llama). Kelp gulls and dolphin gulls steal eggs, gulls and skuas prey on chicks, and sea-lions and giant-petrels overpower the penguins in the water and tear them to pieces.

Magellanic Penguins suffer from oil spills and the discharge of oily ballast. They are captured in fishing nets, and commercial fishing near colonies reduces food availability. For colonies to be sustained, high adult survival is essential. Therefore conservation efforts should be directed towards those human activities that cause mortality.

# African Penguin *Spheniscus demersus*

## DESCRIPTION

The specific name *demersus* means plunging, or sinking. This species was commonly referred to as the Black-footed or Jackass Penguin, because its feet were black and its call sounded like a jackass or donkey braying. The spheniscids of South America were also given the name of Jackass. The African Penguin's feet are two-toned: black and flesh-coloured.

The African Penguin is the only penguin commonly found in Africa, but three other species, King, Rockhopper and Macaroni Penguins, have been recorded there as rare vagrants, as well as a Gentoo Penguin, a recent ship-assisted arrival.

The adult African Penguin is 68 cm long, its flipper 15–18 cm and its weight 3–3.6 kg. The upperparts are black and the underparts white. A white band reaches from the top of the bill above the eye and around the back of the face, giving the face a mask-like appearance. A single black band runs from the feet, up the flanks to meet across the breast, widening below the flippers. Black spots are scattered on the underparts and sometimes appear to form a second black band above the strong permanent one. During the breeding season bare pink skin surrounds the eye and extends to the bill across the face. The eye is brown, the bill black crossed by a grey band, and the black legs and feet are blotched with pink.

Chicks in second down are brown above with white underparts and usually with a white face. Fledglings are blue-grey on the upperparts and white underneath, with a grey wash on the chin and lower throat, and they lack the black band of the adult.

## DISTRIBUTION, DISPERSAL AND POPULATION

African Penguins live only in southern African coastal waters. The most populous region is that served by the Agulhas Current. Colonies are distributed on offshore islands and, rarely, on the mainland from Algoa Bay in south-east South Africa to Hollams Bird Island, Namibia, on the west coast.

Most African Penguins do not go further than 12 km from land. Stragglers, mainly young, have been recorded as far away as Mozambique and Gabon. In one recorded movement an African Penguin travelled over 700 km at an average speed of 30 km/day and another one, rehabilitated after being oiled, travelled 900 km at an average speed of 81 km/day.

Around the turn of the century the population was thought to number more than one million birds, but it had decreased to an estimated 160 000 birds in 1993. In the 'South African Red Data Book — Birds' (Brooke 1984), the African Penguin is listed as 'vulnerable' and from Dassen Island northwards it is considered 'endangered'. Despite this trend, there are examples of successful recolonisation. Birds contaminated by oil spills and rehabilitated after cleaning were released on Robben Island near Cape Town, an island devoid of penguins for 180 years. Nearly 3000 birds had been released on this 500 ha island before it was recolonised by birds from other breeding colonies. Beginning with nine pairs in 1983, the population increased to more than 2200 pairs in 1993. Two other new breeding colonies became established on the mainland in the same period. It is likely that most colonists were young first-breeders that emigrated from other breeding colonies as a result of food scarcity in their vicinity.

## AT SEA AND ON LAND

The African Penguin forages by day close to its home islands in small groups of fewer than ten birds, but rafts of about 3000 have been recorded at sea. Birds act in unison when swimming and diving. When groups are larger than ten or so, actions are not in unison. Birds submerge for short periods and feed in association with cormorants, gannets and terns. Birds on a 24-hour foraging trip covered an average of 102 km, at an average distance of 39 km from their island. Some parents feeding chicks leave after dark and swim mainly on the surface to reach their feeding area at daybreak. On their return they swim mainly under water.

When entering the water they walk in, or else jump feet first from rocks.

Birds not involved in breeding may spend the day at sea and return to their islands at night, thus avoiding the heat of the day ashore.

## BEHAVIOUR

African Penguins are gregarious and their colonial display is ritualised. The following abbreviated description of their display sequences is taken from *The Birds of Africa* (Brown, Urban & Newman 1982). In the Ecstatic Display, indicative of the ownership of territory or advertising for a mate, the male, either alone or in company with his mate, stands and slowly stretches his head directly upwards, raising his flippers sideways. Opening his bill, he throbs silently, then immediately breaks into the full braying call. This display may be performed and repeated for an hour or more, others in the colony sometimes joining in.

Females may bow when approaching another bird. If either sex wants to attract the other, the head is arched downwards and the bill either pointed to the ground or at an angle towards the other. They sidle around each other and the male, returning to the nest site, crouches and vibrates his head from side to side, giving a symbolic scrape of the foot as if hollowing out the nest. If the female is receptive, she also crouches and quivers her body. This is followed by the Arms Act where the male presses against the female's head and back if she is standing, patting her with his flippers until she subsides and copulation follows. If she is unreceptive she will move away or, if she is already crouched, she will topple the male from her back.

There are variations on the Ecstatic Display which include

bowing between a pair with the head held in positions from the vertical to the horizontal.

In aggressive encounters birds may stretch to place their bills side by side, then vibrate their heads rapidly so that their bills slap together. They peck, interlock bills, and pull and twist, trying to down their opponents. They will bite an offending individual and, while holding on, wallop it with one or both flippers until the offender escapes.

### BREEDING

The breeding season varies from place to place, but laying is synchronised within each colony. Laying may begin in January (summer) and continue through to October by replacement laying after the failure or success of the first clutch. A second clutch may also be raised successfully. Sometimes a third replacement clutch follows two failures and occasionally even a fourth clutch may be laid as late as September, but this is rarely successful. Some birds do not breed when food is scarce.

Nests are built in randomly spaced burrows in accumulated guano deposits or impacted soil, beneath and between rocks, artificial structures or large items of jetsam. When suitable burrowing surfaces are not available, nests are built above ground from vegetation, feathers, stones or debris and are closely packed next to one another pecking distance apart. Birds steal material from other nests.

When food is plentiful one parent guards the chicks while the other forages. Young chicks that are left unguarded at the nest are killed by kelp gulls. Older unguarded chicks form crèches at about four weeks and huddle together for warmth and protection. Crèching is more common in chicks raised on the surface. They go back to their nests in the late evening to await their parents' return from the sea.

Fledglings leave the colony at about 12 weeks and may return as juveniles to their natal area after a year, but some do so much later. After long periods at sea they have obvious difficulty in balancing when standing and walking at their first return.

They first breed at three to four years. Some African Penguins are known to have lived longer than 22 years.

Although the availability of sardines may cause the birds to nest, breeding success is closely matched to the availability of anchovies.

African Penguins moult once annually over a period of 21 days on their home islands, on beaches or at nest sites. Moult may take place at any time of the year, but generally there are two peaks following breeding. Incompletely moulted birds that enter the water because they are starving do not survive.

## THREATS AND CONSERVATION

In the seventeenth and eighteenth centuries, penguins on Robben Island were clubbed or shot in large numbers for food and oil, resulting in their absence for nearly two centuries. Commercial collection of guano removed material in which the penguins could dig burrows, causing some penguins to become surface nesters. Unshaded nests are susceptible to desertion when the parents are stressed by heat. Legal harvesting of eggs ceased by 1968, but people still steal them. People unwittingly cause harm when they approach the penguins too closely. This causes panic and temporary nest desertion which allows the kelp gulls to prey on eggs and chicks. The mole snake on Robben Island is a natural predator on eggs and possibly small chicks, and parasites can cause high mortality. Introduced predators, such as feral cats, prey heavily on chicks. On one island the construction of a causeway gave mammals access to a colony, but predation has been reduced by the erection of a wall. The mongoose is a predator at mainland colonies.

Oil pollution is a continuing threat, but the efforts of the South African National Foundation for the Conservation of Coastal Birds (SANCCOB) in particular, have ensured that there is a high survival rate of contaminated penguins, which are then returned to sea. The Robben Island colony has continued to grow despite the rise in the number of tourists under supervision of guides; this lack of harmful impact is encouraging. There have been some fatalities from traffic, but speed humps and a detour road have reduced the danger.

The reasons for penguins choosing to recolonise Robben Island were the availability of breeding space, sufficient density of food, the presence of large numbers of seabirds (cormorants, gulls and terns) and the availability of shade.

The periodic reduction of food fish may be caused by natural fluctuations, as well as by over-exploitation of fish stocks. Unusual oceanographic events in 1982–84 (the ENSO phenomenon) may have caused a decline in the population.

Conservation action is aimed at protecting breeding colonies by removing feral cats and, at particular colonies, by restricting the spread of fur seals. Fur seals consume fish nearby and also effectively exclude the sharing of breeding space. To protect penguins from seals, which kill and eat them, ten asbestos pipes were put in place on Seal Island (False Bay) and within a month nine were occupied by penguins. Further conservation effort needs to be directed towards reducing oil pollution, controlling guano harvesting and monitoring seal numbers at penguin colonies.

# Humboldt Penguin *Spheniscus humboldti*

## DESCRIPTION

These penguins are so named because they rely on the Humboldt Current for their food. They are probably the least studied of all penguins in the wild, despite the fact that they live in accessible places in a temperate climate, unlike the much-studied Antarctic species that live in inaccessible places with a hostile climate. The fact that they are shyer than most other species may account for this lack of study.

At 67–72 cm long, with flippers of 16–17.5 cm and weighing about 4 kg, they resemble both the African and Magellanic Penguins, but have only one black breast band. The eye is reddish-brown. The bill, which is black with a few lighter spots, is heavier and longer than that of the Magellanic Penguin. The feet are blackish-brown.

Chicks in second down are brown above and white below, and their eyes are grey. The heads of fledgling Humboldt Penguins are brown in tone, distinguishing them from Magellanic Penguins which are grey. The spots of juvenile Humboldt Penguins can change during moult into adult plumage and the iris may change from pale to deep red by four to five years of age, though some old birds have dark eyes.

## DISTRIBUTION, DISPERSAL AND POPULATION

Humboldt Penguins live almost exclusively in a long narrow band of coastal water where the Humboldt Current is in contact with the coasts of Peru and Chile and the islands bordering them. It is a region without fresh water or rain. The Humboldt Penguin's

131

range overlaps that of the Magellanic Penguin by more than 300 km. The two species normally breed in separate colonies, but they occasionally interbreed. When not breeding, Humboldt Penguins and Magellanic Penguins share the same seas for a distance of about 1000 km. The population has declined greatly and in 1993 there were thought to be fewer than 10 000 birds.

### AT SEA AND ON LAND
Apart from the usual foraging pursuit dive, Humboldt Penguins have been observed to hunt among the barnacles and weed attached to the hulls of ships. Humboldt Penguins often bathe immediately they enter the water. Some nights they call continuously. If sea-lions lie piled up across the beach, Humboldt Penguins walk between and even over them. Sea-lions do not appear to cause fear, even though they are known to attack the similar Magellanic Penguin.

### BEHAVIOUR
Humboldt Penguins are gregarious; on land they are timid. Their displays appear to be similar to those of other spheniscids. Very occasionally, Humboldt Penguins have been recorded with their heads tucked under a flipper when asleep.

### BREEDING
Males arrive at the breeding area a few days before females and are the only ones to carry nesting material. Eggs are laid from the end of February into March, but it appears the Humboldt Penguin may breed at any time of the year and is capable of rearing two clutches in succession. Because most of the guano in which they previously dug their burrows has been removed, they have had to resort to caves in precipitous cliffs well away from the water. Pelican feathers are among the materials used for nesting.

Humboldt Penguins moult annually on the beach, not in the nesting area; the time of the year varies, as does the breeding season.

## THREATS AND CONSERVATION

Humboldt Penguin numbers have declined because the Humboldt Current is overfished. The birds are eaten, used as fish bait and caught in nets. Stealing eggs has also reduced the population. The deep guano built up on their breeding islands has been harvested for its rich fertiliser properties. Where it has been scraped off down to the bare rock there is nowhere for Humboldt Penguins to dig their burrows and consequently, even if they attempt to breed, they fail. Wild dogs and other predatory mammals such as foxes and cats bar them from breeding on most of the mainland coast.

Conversely, the erection of predator-proof walls on headlands of the mainland in Peru has led to increases in the population of Humboldt Penguins.

# Galapagos Penguin *Spheniscus mendiculus*

## DESCRIPTION

The species name *mendiculus* means beggarly, which refers to the small size of the Galapagos Penguin compared to other penguins.

With a length of 53 cm, a flipper of 11–12 cm and a weight of about 2–2.5 kg, the Galapagos Penguin is the smallest of the four spheniscids. It is never likely to be mistaken for the other three in its wild state because it is found only in the Galapagos Islands. The back is black, the belly white and flecked with black feathers. A narrow white line stretches from the pinkish eye to the side of the throat and a grey-black band across the breast is indistinct. The long and slender bill is black with the lower part yellowish. Feathers under the bill form a white chin patch. Bare skin at the base of the bill and around the eye is pink with black splotches during breeding. The amount of bare skin and the pattern of colour varies between individuals and can be used for recognition, as can the belly spots. The colouring of males is more definite than that of females. Feet are black with white modelling. Chicks in second down are brown above and white underneath. Fledglings have grey backs and white cheek patches and they lack the adult breast and head bands and the lighter colour on the bill.

## DISTRIBUTION, DISPERSAL AND POPULATION

The Galapagos Penguin is the most truly warm-weather penguin, breeding as it does on hot desert islands right on the equator where the air temperature may exceed 40°C and the surface water temperature may vary as much as 14° to a high of nearly 29°C. Even so, like all other penguins, it feeds in cool water, the cool

Cromwell Current providing suitable conditions in the Galapagos Islands. Breeding is mainly along part of the coasts of Fernandina and Isabela Islands.

The Galapagos Penguin does not disperse away from the Galapagos Islands, and adults and juveniles are ashore at their breeding locations throughout much of the year. Juveniles often associate with adults during the breeding season, but the different colouration of juveniles may reduce any adult aggressive or pair-bonding attempts.

The population was estimated in 1970–71 to be between 6000 and 15 000 birds. The population fluctuates depending on oceanic changes and it is particularly low after El Niño events.

## AT SEA AND ON LAND

Galapagos Penguins forage by day, usually solitarily or in small groups, but groups of 200 penguins have been observed when small fish are abundant. Small schooling fish are more prevalent when surface waters are cool.

The penguins' dark backs absorb warmth. They can either warm themselves by basking in the sun or cool off by swimming in the ocean. In surface swimming the flippers remain submerged, describing a small arc unlike the action used under water. They seldom porpoise and often return to land to sleep.

When walking they do not sway or have the flippers extended unless hurrying. They seldom attempt to toboggan because of the rough surface of the ground.

## BEHAVIOUR

Galapagos Penguins have adapted to living in extremes of temperature. They forage in the ocean during the day, but spend the night on land. They cannot escape the cold of the water when foraging nor the heat on land when in the open, though they seek shade by nesting in crevices. On land they stand hunched over to shade their feet, with the flippers held out to shade their undersides, taking the full effect of the sun on their backs. They lose heat by rapid panting (the hotter they get the faster they pant) and also from the extremities (the feet and undersides of the flippers), which are not as heavily insulated as the body. Non-breeders or those off-duty that are not forced to stay on land will jump into the water when hot.

Breeding birds display at their nests or in the water, their displays being similar to other members of the group. Bill Duelling is a ritual displayed by Galapagos Penguins when a bird leaves the water and approaches another closely. The two may be either mated or unmated. They face each other, shake their heads and clatter the tips of their bills together. Another ritual, also when two birds move close together either in the water or on land, is Head Movement, which may be a greeting ceremony or an appeasement response to threat. The bill and head are swung in an arc that begins on one side of the body, then swings up and around to the other side. The intruder performs the same movement, both birds repeating it until either the intruder moves away or the defender pecks at it. Copulation takes place both on land and at sea.

Galapagos Penguins sleep lying prone, with the flippers tucked under or alongside their bodies; rarely is the bill tucked under a flipper.

## BREEDING

Breeding has been recorded in every month of the year, but it is more common when the water temperature is below 24°C. When the water temperature rises, the nutrients on which small fish feed are lacking. Consequently the small fish on which penguins prey are too scarce and the birds have more difficulty finding food. When food is scarce breeding does not begin, and if already underway it ceases. The lower the temperature, the greater the breeding success. Nests are shaded in crevices, lava fissures and tunnels, and there are two or three clutches per pair during the year.

Breeding takes place when there is an abundance of food which can be taken advantage of quickly by those with long-lasting pair bonds. Those who must go through the ritual of courtship may find that prey is no longer abundant when they need it for chick rearing.

Adults must remain in attendance at the nest until chicks are about four weeks old to protect them from predators.

The Galapagos Penguin is unusual in that it may moult twice a year, each time *before* breeding at which time it needs unimpaired plumage to insulate it from the intense heat and protect it from the cool water. It breeds in response to an abundance of food, and *after* breeding it may not be assured of sufficient food to complete its moult. Moulting birds avoid the water, but the heavy fat deposits laid down as a prelude to moult may help protect them from overheating. Some Galapagos Penguins do enter the water before completing moult, if they have low body weights and need to forage to avoid starvation, which is the greatest cause of mortality.

The life expectancy of Galapagos Penguins is unknown, but three breeding birds were known to be over 11 years of age.

## THREATS AND CONSERVATION

In the years when El Niño reaches the Galapagos Islands and the mean surface water temperature rises to 25°C, Galapagos Penguins do not breed. Warm surface waters lack food and therefore lead to breeding failure. The Galapagos Penguins had not recovered in number three years after the 1982–83 ENSO event, the severest recorded period of warm ocean temperatures and heavy rains around the Galapagos Islands.

Decline in the population has also been attributed to predation by feral dogs and cats, capture in fishing nets and to volcanic eruptions. Lava entering the sea raises the water temperature and modifies the coastline around some of the breeding areas.

Crabs, rice rats and snakes are natural predators of young and eggs. Other indigenous predators, such as hawks, owls, sharks and other large sea animals, may prey on penguins but do not significantly alter the size of the population.

## Chapter 8

# Threats, Conservation and the Future

## THREATS AND CONSERVATION

Throughout this book evidence points to the fact that, apart from the natural order of things, the greatest danger to penguins comes from humans. Human populations, along with their introduced pests and pets, have spread into all environments and taken whatever they have wanted without regard for their effect on the environment, its resources or other creatures. Already at risk are the Yellow-eyed, Fiordland and Humboldt Penguins, and the future of the African and Galapagos Penguins is not entirely secure.

We can do something to reduce our impact on penguins, but first we need to know what their needs are and what threatens their survival. That means research. As explained in the first chapter, to obtain accurate biological data necessitates handling penguins and involves some degree of intrusion into their lives. As far as possible, research that causes stress has been reduced with the advent of advanced technology. No matter how much we learn, however, it is doubtful that we shall ever know everything connected with a penguin's existence.

Paradoxically, danger may lie in some conservation methods. For example, where breeding penguin calls have been recorded and played back, other penguins have been stimulated into laying earlier and more synchronously, which is probably to their benefit. On the other hand, if small declining populations are manipulated to improve breeding success by laying earlier and more synchronously, they may then be at risk from some catastrophic event, such as a rise in sea level, drought or fire. When breeding is asynchronous, at least part of the population survives.

## NATURAL ENEMIES

The natural enemies of penguins vary, depending on where they live. There are predators in the sea (sharks, seals, leopard seals, killer whales) and on the land (skuas, giant-petrels, raptors, ravens, snakes). We can do nothing about natural predators, nor should we wish to do so, because a balance, no matter how uneasy, is necessary between prey and predator. Predators that occur naturally help to keep populations under control: otherwise, many penguin species might eat themselves out of existence. It is possible that the apparent rise in Antarctic penguin populations can be attributed to the reduction in whale populations with which penguins compete for food.

Penguins themselves are predators. They feed on organisms that live in the sea. These too need to be maintained in some sort of balance, but natural phenomena sometimes throw the system out of balance. The most nutrient-rich waters are those where cool currents rise. In this upwelling water live the prey of penguins: fish, krill and squid. Periodically these cool currents are displaced by warmer ones, the best known of which is El Niño (associated with the Southern Oscillations as ENSO events) on the west coast of South America. The effect of this is widespread, resulting in an intense shortage of food for the seabirds of the Pacific coast, including penguins. They may either fail to breed or, if they have begun, be unable to feed their young. About these ENSO events, which have their repercussions in the southern parts of Africa and Australia, we can do nothing.

## HUMAN INTERFERENCE

Because we have no control over natural events, it is even more important that we do control those other events that result from past ignorance and present greed.

Figs 53 and 54. *Right* Undisturbed Gentoo Penguins.

In temperate coastal areas, our desire to farm, to harvest forests and to live near the sea has resulted in the alienation of penguin habitat. This has been particularly drastic for the Yellow-eyed Penguin in New Zealand, which has the smallest population of all the penguins. Attempts are being made to halt this decline, but both replacement of natural habitat and protection from introduced predators are difficult to achieve. When a small non-migratory population is confined to small isolated colonies and interbreeding between the colonies does not take place, it is at risk from natural catastrophes. Even the most abundant species are at risk if their habitat and their food supplies are destroyed.

On Phillip Island in Victoria, similar habitat reduction had occurred, but this has not only been halted by the prohibition of further encroachment into Little Penguin breeding grounds, but has actually been reversed by buying back land and regenerating it for penguins, by controlling visitors who wish to view the penguins, by attempting to eradicate feral and introduced animals and by controlling traffic. The program to implement all this is quite recent, and perhaps it is too early to claim success — but it is encouraging.

On Macquarie Island, sealers used to extract oil from seals in big boilers. When the supply of these ran low, they turned to King Penguins and almost completely exterminated them. King Penguins are trusting animals and it was easy for the sealers to herd them on to ramps that led to the top of boilers where the birds fell in. The subsequent banning of both this practice and the use of their eggs and flesh for food and their feathers for trimmings on clothing, has led to King Penguins now breeding there again in large numbers.

Harvesting of guano in temperate regions has removed the surface material in which penguins dig their burrows, thereby reducing available habitat. The damage caused has been recognised and arrested in South Africa, but it is doubtful that the same applies where Humboldt Penguins breed in South America.

The most abundant predator at the top of the food chain can legitimately claim a proportion of the production of the oceans. The human is that predator. However, when commercial fisheries use long lines and vast lengths of drift nets (perhaps analagous to the clear-felling of forests), we need to consider the implications. Many non-target species are drowned: albatrosses, penguins, dolphins, seals, and non-target fish species are tossed back dead into the sea. Large-scale fishing might be legitimate if the resultant fish harvest were to feed hungry humans. But to feed pets? Or gourmets?

## INTRODUCED ANIMALS

Where humans have intruded into and changed penguin habitat, some species have declined, particularly in temperate regions. Introduced animals (dogs, cats and stoats) prey on penguins, while foxes slaughter birds but do not eat them. Cattle and sheep trample penguin burrows and nests. The elimination of introduced predators is not easy, but where this has been done the results have been dramatic. Without rabbits vegetation grows, and ground-dwelling animals, if not already extinct, begin to increase.

The effect of some predators is more subtle. On some subantarctic islands, cats are not a direct threat to penguins. They prey on burrowing petrels which are also the prey of skuas. Skuas also prey on penguins; when petrel numbers are reduced, their predation on penguins is more intensive. While skuas migrate, cats do not and must consequently feed on whatever is available.

## POLLUTION

Oil pollution, caused not only by tanker spills but also by discharge from bilges, is responsible for much damage to penguins. Penguins cannot detect oil spills and unwittingly swim into them either swimming on the surface or when they come to the surface for air. Oil mats their feathers and allows water to penetrate which causes hypothermia. If they reach the shore, they try to preen the oil from their feathers. This is ingested and its toxicity kills them.

Continuing attention is being given to the rehabilitation of penguins suffering from oil spills, starvation and other injury. If oiled birds are caught as soon as they reach the shore to prevent ingestion of the oil, rehabilitation (after enormous effort by concerned people) is more likely to be successful now that the requirements for housing and treatment are known. The usual practice of penguins is to feed on live prey caught in pursuit diving. They do not recognise dead food, so feeding by hand is necessary. This has been improved by the production of fish sausages: minced fish in sausage skins which can be fed to penguins when suitable sized fish are not available.

Already it has been shown that it is impossible to contain or ameliorate the effect of oil spills from tankers in the Arctic and the Antarctic. The most likely areas for oil exploration in Antarctica would be in the Ross Sea where Adelie Penguins tend to congregate to feed. Test drilling there in 1973 showed supplies of natural gas and indications of oil. Exploration for oil could have a catastrophic effect on penguin populations anywhere in the Antarctic, and it has been expressly prohibited for 50 years by the Environmental Protocol of 1991. That is simply a breathing space. If we care at all for our environment, it is essential that some other form of natural energy be found to replace the need for oil.

Other pollution of the sea also causes damage. Penguins become entangled in discarded fishing line and they also cannot escape when their flippers and legs are imprisoned in the plastic carriers designed to hold cans and bottles. Pieces of plastic regurgitated by parents to chicks limit their stomach capacity and restrict digestion and growth. Chemical pollutants can also cause problems.

We are slowly beginning to appreciate that there are other creatures in the world to be considered, but we have a long way to go in ensuring the protection of even one group that has suffered through our thoughtlessness — penguins.

## HOW TO WATCH PENGUINS

People everywhere are attracted to penguins, and those who visit Antarctica are no exception. Visitors need to be aware of what their presence means to the penguins and to act accordingly.

Research on the Adelie Penguin has revealed the physical stress caused by humans. First, the temperature of Adelie Penguins rises when they are handled and stays high for several hours afterwards. Secondly, the approach of humans to within a few metres of the nest results in a raised heartbeat rate, even though the bird appears unconcerned and shows no outward sign of distress. (An obvious sign of distress is when the penguin flees from the nest, the contents of which are then open to predation. When one bird flees, others are likely to follow.) When approached by a natural predator, for example a sheathbill, the rise in the heartbeat is not as high.

As recently as 1992, helicopters were permitted to approach to within 200 m of a nesting colony of Adelie Penguins. Early in the season adults stayed on their nests but later, when chicks crèched, adults and chicks fled from the colony when the helicopter was still 300 m away. Later still, even with the helicopter 1500 m away, chicks scattered, having by then become aware of the potential danger. As a consequence of disturbance to birds undergoing long fasts during nesting (and moulting), energy reserves may be depleted rather than conserved, increasing the chances of breeding failure. Disturbance by humans needs to be kept to a minimum and greater restrictions placed on the approach of helicopters.

A bird previously handled may react more strongly to the approach of a human than one not previously exposed. On the other hand, according to some researchers, if humans restrict their approach distance to a minimum of 30 m Adelie Penguins may not react because they do not recognise them as large land-based predators of which there are none in the Antarctic.

Fig. 55. Nesting sites such as this one are highly sensitive to disturbance by humans.

*All visitors to penguin breeding sites need to understand that a common sign of fear is the cessation of activity even though the bird remains on the nest.*

African Penguins standing on beaches did not appear disturbed by the approach of one person who stopped at intervals and did not approach closer than 10 m, but a slow steady approach caused greater disturbance.

Where crowds are controlled, such as at the Penguin Reserve on Phillip Island, penguins are not under threat from watchers provided the latter observe the rules laid down at the reserve.

## THE FUTURE

In any prediction, uncertainty is always present and this applies to predictions as to what may happen to penguins in Antarctica. One thing is certain: there will be change.

Change has been constant throughout the world's history, as is revealed by the fossil record and examination of ice cores. Within the large-scale changes in the Antarctic environment over the last few million years are dramatic short-term fluctuations of climate. These short-term fluctuations, cold or warm snaps that deviate from the long-term mean, may explain periods in which many species of marine fauna became extinct. Without such fluctuations, species would have been able to adapt to the very slow rates of cooling or warming of sea water during long-term changes of temperature.

Changes in much of coastal Antarctica will be noticeable early in the twenty-first century, but ideas conflict as to which way that change will go. One hypothesis suggests that the greenhouse effect will cause the world to become warmer over the next 50–100 years, but it is also possible that the world will enter a new ice age within a few thousand years due to predicted changes in the earth's orbit around the sun.

Examples of changes in the southern oceans are evident. In the Antarctic on Signy Island, summer (but not as yet winter) air temperatures have risen in recent years, with a consequent reduction in the area of its permanent ice-fields. Similarly, for nearly three years from mid-1988, the extent of the sea-ice to the west of the Antarctic Peninsula has shown a major decrease which coincides with rising air temperatures. In the subantarctic region on Campbell Island, the drastic reduction of 94 per cent of the Rockhopper population appears to be associated with rising sea surface temperatures. From a mean of 9.1°C in 1944, temperatures have fluctuated both up and down to settle on a mean of 9.7°C since 1970. Elephant seal and albatross populations on Campbell Island have also declined drastically. Although figures are not known precisely, this trend towards regional warming also appears to be taking place at the Antipodes and Auckland Islands.

These are but a few examples of warming and melting, and it cannot necessarily be inferred that the same conditions apply to the more southerly mass of the Antarctic continent. These changes may have arisen as a result of naturally occurring fluctuations over a much longer term and time will show whether there is a trend.

Another potential but as yet undocumented threat to penguins over the next several decades is the depletion of ozone in the upper atmosphere, which is largely due to increases in chlorine concentrations from chlorofluorocarbons (CFCs). This ozone depletion is greatest over Antarctica in the southern spring and early summer. Total ozone amounts over the Antarctic coastal zone have decreased by up to 30 per cent in the spring months since the 1960s and 1970s, and by up to 12 per cent as far north as Macquarie Island. Smaller decreases of about 5 per cent have occurred — or are expected to occur — over the next decade even at the latitude of Hobart, Tasmania.

Biologically damaging ultraviolet (UV) radiation generally increases by about twice the percentage decrease in ozone. UV radiation is thus expected to be as great over Antarctic waters as in India where, unlike the Southern Ocean, plants and animals have evolved to cope with the higher levels. This increase in damaging UV radiation has now been measured in Antarctic waters and is found already to be causing a reduction in primary marine production. This might well be expected to affect penguins which lie further up in the food web that starts with the phytoplankton and Antarctic krill.

Fortunately, compliance with the Montreal Protocol on the Protection of the Ozone Layer seems to have greatly reduced the rate of increase in chlorine concentrations in the upper atmosphere. This is expected to lead to a levelling off of the ozone depletion

Fig. 56. *Right* King Penguins and elephant seals at Macquarie Island.

by about 2010, with a subsequent slow recovery over the next century. How penguin populations will be affected in the meantime is uncertain. We can, however, do something to influence our future climate; we can alter our behaviour to reduce our profligate use of material known to pollute the atmosphere.

If temperature change is responsible for extinctions, then its direction is of no consequence. Excessive heat or cold are equal problems.

How such extreme temperature changes will affect penguins we can only hypothesise. If marine resources become so depleted that they cause the extinction of penguins, will humans, at the top of the food chain, go the same way, or will we be better able to adapt?

By looking at just one family of animals — penguins — in different parts of the southern hemisphere, it is evident that the environment of the world is not divided into separate compartments. It is a complete whole and we are only a small part of it, albeit with the ability to wreck or preserve. Despite the emerging evidence, there is room for optimism: conceivably the scientists might just have got it wrong.

Fig. 57. *Right* Anne Kerle, who joined the author in a study of Gentoo Penguins, awaits their entry in a corral. This method causes less stress than capturing the birds after a chase.

# Glossary

**allopreen**   Preening of one bird by another.

**amphipods**   Small crustaceans.

**Antarctic Convergence**   More usually called Antarctic Polar Front.

**Antarctic Polar Front**   Circumpolar zone where cold Antarctic surface-water sinks below less dense subantarctic surface-water; northernmost extent coincides with 2°C subsurface isotherm.

**antiphonal**   One bird gives a call and a second one answers it, often with a shorter call.

**axilla**   The 'armpit'; underneath the junction of wing or flipper and body.

**cape anchovy**   *Engraulis capensis*.

**caracara (falcon)**   *Phalcoboenus* sp.

**cephalopod**   Squid.

**cloaca**   The external opening for both the alimentary tract and reproductive system.

**clutch**   The set of eggs laid during one breeding episode.

**crèche**   An assembly of dependent young in the nesting area.

**crustacean**   Small shrimp-like marine animals.

**dabbling**   One bird bends down and is imitated by its partner.

**elephant seal**   *Mirounga leonina*.

**El Niño**   Climatic event named after the child Jesus because it occurs around Christmas time and was regarded as a time of plenty on land in South America following abundant rain. El Niña is the reverse of El Niño.

**ENSO (El Niño/Southern Oscillation)**   The cold current that flows northward along the coast of Peru becomes displaced by the warmer southward-flowing Pacific Equatorial Counter-Current. Upwelling of cold water carrying nutrients ceases with a consequent lack of food for fish, the prey of many penguins. The ENSO event of the greatest severity occurred in 1982–83 affecting penguins in the Galapagos Islands and in South Africa, and producing severe droughts and catastrophic bush-fires in south-eastern Australia.

**eudyptids**   The *Eudyptes* genus, the crested penguins.

**fairy prion**   *Pachyptila turtur.*

**ferret**   *Mustela putorius.*

**fox**   *Vulpes vulpes.*

**fur seal**   *Arctocephalus* spp.

**genus**   Group (plural genera).

**giant-petrel**   *Macronectes* sp.

**gravid**   Pregnant, egg-bearing.

**guanaco**   Llama (*Lama huanacos*).

**guano**   Bird excrement.

**gull**   *Larus* spp.

**higher latitudes**   Colder regions closer to the poles.

**hypothermia**   The condition of having an abnormally low body temperature.

**isotherm**   A line on a map joining places with the same temperature.

**killer whale**   *Orcina orca.*

**krill**   Crustaceans, euphausiids.

**leopard seal**   *Hydrurga leptonyx.*

**lower latitudes**   Warmer regions away from the poles.

**mass**   Weight.

**moult**   The shedding and replacement of feathers, usually once annually.

**occipital**   The back of the crown.

**pelagic**   At sea far from land.

**phytoplankton**   Plankton consisting of plants.

**plankton**   Small organisms that float in the sea.

**polynya**   Open water surrounded by ice.

**pre-egg**   Before egg-laying.

**pygoscelids**   The genus *Pygoscelis*, the stiff-tailed penguins.

**rice rat**   *Oryzomys nesoryzomis*.

**rookery**   A colony; a group of animals breeding in one place. Preferably reserved for rooks (a European bird).

**sardine**   *Sardinops* sp.

**scrape**   A depression formed in the ground to make a nest.

**sea-eagle**   *Haliaeetus leucogaster*.

**seal**   *Arctocephalus* sp.

**sea-lion**   *Otaria* and *Neophoca* spp.

**sedentary**   Abiding in one place, not migratory.

**shearwater**   *Puffinus* spp.

**sheathbill**   *Chionis* sp.

**skua**   *Catharacta* spp.

**skua, subantarctic**   *Catharacta lonnbergi*.

**spheniscids**   The genus *Spheniscus*, the penguins with black chest bands.

**squid**   Cephalopod.

**stoat**   *Mustela erminea*.

**subantarctic zone**   Between the temperate and Antarctic zones.

**superciliary**   Eyebrow.

**synchronous**   At the same time.

**tarsi (plural tarsus)**   The fused and elongated foot bones that look like the lower leg.

**temperate zone**   Between the equatorial and subantarctic zones.

**toboganning**   The penguin slides along on its belly, sometimes propelling itself with flippers and feet.

**vagrant**   A bird that wanders outside the usual range.

**weka**   *Gallirallus australis*.

# Bibliography

To save space, where a number of papers have been referred to from the one source they are listed alphabetically under the parent publication.

Ainley, D. G. 1980. Survival and mortality in a population of Adélie penguins. *Ecology* 61: 522-530.

Bengston, J. L., Croll, D. A. & Goebel, M. E. 1993. Diving behaviour of Chinstrap penguins at Seal Island. *Antarctic Science* 5: 9-16.

Boersma, P. Dee. 1976. An ecological and behavioral study of the Galapagos penguin. *Living Bird* 15: 43-93.

Boersma, P. D. 1978. Breeding patterns of Galapagos penguins as an indicator of oceanographic conditions. *Science,* vol. 200: 1481-83.

Brooke, R. K. 1984. South African Red Data Book—Birds. *Rep S Afr Natn Scient Progrm* 97: 1–213.

Brown, L. H., Urban, E. K. & Newman, K. 1982. *The Birds of Africa* vol. 1. Academic Press. 76-80.

Cannell, B. L. 1992. Feeding behaviour of Little penguins *Eudyptula minor* in captivity. *Corella* 16: 139.

Clark, J. R. & Kerry, K. R. 1992. Foraging ranges of Adélie penguins as determined by satellite tracking. *Corella* 16: 140.

Crawford, R. J. M., Williams, A. J., Randall, R. M., Randall, B. M., Berruti. A. & Ross, J. B. 1990. Recent population trends of Jackass penguins *Spheniscus demersus* off Southern Africa. *Biological Conservation* 52: 229-243.

Croxall, J. P., Briggs, D. R., Kato, A., Naito, Y., Watanuki, Y. &

Williams, T. D. 1993. Diving pattern and performance in the Macaroni penguin *Eudyptes chrysolophus*. *J. Zool. Lond.* 230: 31-47.

Cullen, J. M., & Whitehead, M. 1992. Some aspects of the diving behaviour of Adèlie penguins *Pygoscelis adéliae*. *Corella* 16: 141.

Cunningham, D. M. & Moors, P. J. 1994. The decline of Rockhopper penguins *Eudyptes chrysocome* at Campbell Island, Southern Ocean, and the influence of rising sea temperatures. *Emu* 94: 27–36.

Dann, P., Norman, I. & Reilly, P. N. (eds). *Penguin Biology: Advances in research and management.* (in press). Surrey Beatty. Australia.

Crawford, R. J. M., v.d. Boonstra, H. G., Dyer, B. M. & Upfold, L. The recolonization of Robben Island by African penguins *Spheniscus demersus*.

Croxall, J. P., & Rothery, P. Interannual variation in population size and reproductive performance in Gentoo and Macaroni penguins at South Georgia.

Gibbs, P. J. Heavy metal and organochlorine concentrations in tissues of the Little penguin *Eudyptula minor*.

Kerry, K., Clarke, J. & Else, G. The foraging range of Adélie penguins at Bechervaise Island, MacRobertson Land, Antarctica, as determined by satellite telemetry.

Le Maho, Y. How to study breeding penguins without human disturbance.

Nimon, A. J. & Stonehouse, B. Penguin responses to humans in Antarctica: some issues and problems in determining disturbance caused by tourist parties.

Waas, J. R. Social stimulation and reproductive schedules: does the acoustic environment influence the egg-laying schedule in penguin colonies?

Wieneke, B. C., Wooller, R. D. & Klomp, N. I. The ecology and management of Little penguins on Penguin Island, Western Australia.

Wilson, R. & Wilson, M. The foraging ecology of the African penguin *Spheniscus demersus*.

Davis, L. S. & Darby, John T. (eds). 1990. *Penguin Biology*. Academic Press, San Diego, Ca.

Boersma, P. D., Stokes, D. L. & Yorio, P. M. 1990. Reproductive variability and historical change of Magellanic penguins (*Spheniscus magellanicus*) at Punta Tombo, Argentina. 15–43.

Bost, C. A. & Jouventin, P. 1990. Evolutionary ecology of Gentoo penguins (*Pygoscelis papua*). 85–112.

Cockrem, J. F. 1990. Circadian rhythms in Antarctic penguins. 319–344.

Cooper, J., Brown, C. R., Gales, R. P., Hindell, M. A., Klages, N. T. W., Moors, P. J., Pemberton, D., Ridoux, V., Thompson, K. R. & Van Heezik, Y. M. 1990. Diets and dietary segregation of crested penguins (*Eudyptes*). 131–156

Croxall, J. P. & Davis, R. W. 1990. Metabolic rate and foraging behaviour of *Pygoscelis* and *Eudyptes* penguins at sea. 207–228.

Dann, P. & Cullen, J. M. 1990. Survival, patterns of reproduction and lifetime reproductive output in Little blue penguins (*Eudyptula minor*) on Phillip Island, Victoria, Australia. 63–84.

Darby J. T. & Seddon P. J. 1990. Breeding biology of Yellow-eyed penguins (*Megadyptes antipodes*). 45–62.

Davis, L. S. & Speirs, E .A. H. 1990. Mate choice in penguins. 377–397.

Fordyce, R. E. & Jones, C. M. 1990. Penguin history and new fossil material from New Zealand. 419–446.

Kooyman, G. L. & Ponganis, P. J. 1990. Behaviour and physiology of diving in Emperor and King penguins. 229–242.

Lamey, T. C. 1990. Hatch asynchrony and brood reduction in penguins. 399–416.

Sadlier, R. M. F., & Lay, K. M. 1990. Foraging movements of Adélie penguins (*Pygoscelis adéliae*) in McMurdo Sound. 157-179.

Trivelpiece, W. Z. & Trivelpiece, S. G. 1990. Courtship period of Adélie, Gentoo and Chinstrap penguins. 113-127.

Waas, J. R. 1990. An analysis of communication during the aggressive interaction of Little blue penguins (*Eudyptula minor*). 345-376.

Wilson, R. P. & Wilson, M. P. T. 1990. Foraging ecology of breeding spheniscus penguins. 181-206.

Duffy D. C. 1983. Competition for nesting space among Peruvian guano birds. *Auk* 100: 680-688.

El-Sayed, S. Z. & Stephens, F. C. 1992. Potential effects of increased ultraviolet radiation on the productivity of the Southern Ocean. In Dunnett, D. A. & O'Brien, R. J. (eds). *The Science of Global Change: The Impact of Human Activities on the Environment*. ACS, Washington, DC. 188-206.

Emlen, J. T. & Penney, R. L. 1964. Distance Navigation in the Adèlie penguin. *Ibis* 106: 417-431.

Gendner, J. P., Challet, E., Handrich, Y., & Le Maho, Y. 1992. Automatic weighing of electronically-identified King penguins. *Corella* 16: 144.

Harrison, P. 1983. *Seabirds: An Identification Guide*. Croom Helm, Kent, UK: 207.

Jacobs, S. S. & Comiso, J. C. 1993. A recent sea-ice retreat west of the Antarctic Peninsula. *Geophysical Research Letters* 20, no. 12: 1171–74.

Kerry, K., Clarke, J. & Else, G. 1992. The use of an automated weighing and recording system for the study of the biology of Adélie penguins. *Proc NIPR Symp Polar Biol* 6: 62–75.

Kerry, K. R. & Hempel, G. (eds). 1990. *Antarctic Ecosystems. Ecological Change and Conservation*. Springer-Verlag, Berlin.

Clark, A. 1990. Temperature and evolution: Southern Ocean cooling and the Antarctic marine fauna. 9–22.

Culik, B., Adelung, D. and Woakes, A. J. 1990. The effect of disturbance on the heart rate and behaviour of Adélie penguins (*Pygoscelis adéliae*) during the breeding season. 177–182.

Davis, L. S. & Miller, G. D. 1990. Foraging patterns of Adélie penguins during the incubation period. 203–207.

Hunter, S. 1990. The impact of introduced cats on the predator-prey interactions of a Sub-Antarctic avian community. 365-371.

Kooyman, G. L. & Mullins, J. L. 1990. Ross Sea Emperor penguin breeding populations estimated by aerial photography. 169-176.

Quilty, P. G. 1990. Significance of evidence for changes in the Antarctic marine environment over the last 5 million years. 3-8.

Smith, R. I. Lewis. 1990. Signy Island as a paradigm of biological and environmental change in Antarctic terrestrial ecosystems. 32-50.

Trivelpiece, W. Z., Trivelpiece, S. G., Geupel, G. R., Kjelmyr, J. & Volkman, N. J. 1990. Adélie and Chinstrap penguins: their potential as monitors of the Southern Ocean marine ecosystem. 191-202.

Whitehead, M. D., Johnstone, G. W. & Burton, H. R. 1990. Annual fluctuations in productivity and breeding success of Adélie penguins and Fulmarine petrels in Prydz Bay, East Antarctica. 214-223.

Young, E. C. 1990. Long-term stability and human impact in Antarctic skuas and Adélie penguins. 231-236

Klomp, N. I., Meathrel, C. E., Wienecke, B. C. & Wooller, R. D. 1991. Surface nesting by Little penguins on Penguin Island, Western Australia. *Emu* 91: 190–193.

Le Maho, Y., Robin, J. P. & Cherel, Y. 1988. Starvation as a treatment for obesity: the need to conserve body protein. *News in Physical Sciences* 3: 21–24.

The Little Penguin in Victoria: 1992. Supplement to *Emu* 91.

Cullen, J. M., Montague, T. & Hull, C. 1992. Food of Little penguins *Eudyptula minor* in Victoria: comparison of three localities between 1985-1988. 318-341.

Dann, P. 1992. Distribution, population trends and factors influencing the population size of Little penguins *Eudyptula minor* on Phillip Island, Victoria. 263-272.

Dann, P., Cullen, J. M., Thoday, R. & Jessop, R. 1992. Movements and patterns of mortality at sea of Little penguins *Eudyptula minor* from Phillip Island, Victoria. 278-286.

Harrigan, K. E. 1992. Causes of mortality of Little penguins *Eudyptula minor* in Victoria. 273-277.

Hobday, D. K. 1992. Abundance and distribution of pilchard and Australian anchovy as prey species for the Little penguin *Eudyptula minor* at Phillip Island, Victoria. 342-354.

Mickelson, M. J., Dann, P. & Cullen, J. M. 1992. Sea temperature in Bass Strait and breeding success of the Little penguins *Eudyptula minor* at Phillip Island, south-eastern Australia. 355-368.

Norman, F. I. 1992. Counts of Little penguins *Eudyptula minor* in Port Phillip Bay and off southern Phillip Island, Victoria, 1986-1988. 287-301.

Norman, F .I., du Geusclin, P. B. & Dann, P. 1992. The 1986 'wreck' of Little penguins *Eudyptula minor* in western Victoria. 369-376.

Martin, Graham. 1985. Through a penguin's eye. *New Scientist*, 14 March 1985. 29–31.

McLean, I. G., Johns, P. M. & Miskellyt, C. M. 1988. Snares crested penguins: a preliminary life history table. *Cormorant* 16: 130.

McLean, I. G & Russ, R. B. 1992. First survey of Fiordland crested penguins: a review. *Corella* 16: 147.

Meathrel, C. E. & Klomp, N. I. 1990. Predation of Little penguin eggs by King's skinks on Penguin Island, Western Australia. *Corella* 14: 129–130.

Miller, G. D. 1992. Diving patterns of foraging Adélie penguins *Pygoscelis adéliae* near Ross Island, Antarctica. *Corella* 16: 148.

Miller, G. D. 1992. Murder by penguin: Adélie penguins *Pygoscelis adéliae* destroy eggs and kill chicks of the South Polar skua *Catharacta maccormicki*. *Corella* 16: 147.

Murphy, Robert Cushman. 1936. *Oceanic Birds of South America*, vol. 1. Macmillan, New York. 437, 453, 466.

Obendorf, D. L. & McColl, K. 1980. Mortality in Little penguins *Eudyptula minor* along the coast of Victoria. *Journal of Wildlife Diseases* 16: 251–259.

Philander, G. 1989. El Niño and La Niña. *American Scientist* 77, no. 5: 451-459.

Pittock, A. B. 1993. The consequence for Australia and our region of global climate change and ozone depletion: Australia's responsibility to limit environmental damage. CSIRO paper presented at MAPW National Conference, Lorne, 27 Feb. 1993.

Randall, R. M. 1989. Jackass Penguins. IN Payne, A. I. L. & Crawford, R. J. M. (eds). *Oceans of Life off Southern Africa*. Vlaeberg, Cape Town. 244-56.

Robertson, G. 1990. Huddles. *Aust Geographic* 20: 74-97.

Robertson, G. 1992. Population size and breeding sucess of Emperor penguins *Aptenodytes forsteri* at Auster and Taylor Glacier colonies, Mawson Coast, Antarctica. *Emu* 92: 65-71.

Scholten, C. J. 1987. Breeding biology of the Humboldt penguin *Spheniscus humboldti* at Emmen Zoo. *Int. Zoo Yb.* 26: 198-204.

Scholten, C. J. 1989. Individual recognition of Humboldt penguins. *Spheniscid Penguin Newsletter*, vol. 22: 4-8.

Shelton, P. A., Crawford, R. J. M., Cooper, J., & Brooke R. K. 1984. Distribution, population size and conservation of the Jackass penguin *Spheniscus demersus*. *Sth Afr J Mar Sci* 2: 217-257.

Siegfried, W. R., Frost, P. G. H., Kinahan J. B. & Cooper J. 1975. Social behaviour of Jackass penguins at sea. *Zoologica Africana* 10: 87-100.

Stahel, C. & Gales. R. 1987. *Little Penguin*. NSW University Press, Sydney.

St Clair, C. C. Unfit mothers? Egg ejection by Royal penguin. *Corella* 16: 150.

Stonehouse, B. (ed.). 1975. *The Biology of Penguins*. Macmillan, London.

Boersma, D. 1975. Adaptation of Galapagos penguins for life in two different environments. 101-114.

Boswall, J. & MacIver D. 1975. The Magellanic penguin *Spheniscus magellanicus*. 271-305.

Spellerberg, I. F. 1975. The predators of penguins. 413-434.

Thompson, K. R. 1989. An assessment of the potential for competition between seabirds and fisheries in the Falkland Islands. *Falklands Islands Foundation*: 30-36.

Valle, C. A. & Coulter, M. C. 1987. Present status of the Flightless cormorant, Galapagos penguin and Greater flamingo populations in the Galapagos Islands, Ecuador, after the 1982-83 El Niño. *Condor* 89: 276-281.

van Heezik, Y. M. & Seddon, P. J. 1990. Scrambling for a meal! Competition by selfish Jackass penguin *Spheniscus demersus* siblings. *Corella* 16: 152.

Weavers, Brian W. 1992. Seasonal foraging ranges and travels at sea of Little penguins *Eudyptula minor*, determined by radio-tracking. *Emu* 91:302-317.

Weimerskirch, H., Stahl, J. C. & Jouventin, P. 1992. The breeding biology and population dynamics of King penguins *Aptenodytes patagonica* on the Crozet Islands. *Ibis* 134: 107-117.

Williams, T .D. 1989. Aggression, incubation behaviour and egg-loss in Macaroni penguins, *Eudyptes chrysolophus*, at South Georgia. *Oikos* 55: 19-22.

Williams, T. D. 1990. Annual variation in breeding biology of Gentoo penguins *Pygoscelis papua* at Bird Island, South Georgia. *J Zool Lond* 222: 247-258.

Williams, T. D. 1990. Growth and survival in Macaroni penguin, *Eudyptes chrysolophus*, A- and B-chicks: do females maximise investment in the large B-egg? *Oikos* 59: 349-354.

Williams, T. D. 1991. Foraging ecology and diet of Gentoo penguins *Pygoscelis papua* at South Georgia during winter and an assessment of their winter prey consumption. *Ibis* 133: 3-13.

Williams, T. D. & Croxall, J. P. 1991. Annual variation in breeding biology of Macaroni penguins, *Eudyptes chrysolophus*, at Bird Island, South Georgia. *J Zool Lond* 223: 189-202.

Williams, T. D. & Rothery, P. 1990. Factors affecting variation in foraging and activity patterns of Gentoo penguins (*Pygoscelis papua*) during the breeding season at Bird Island, South Georgia. *Jnl Appl Ecol* 27: 1042-1054

Woehler, E. J. 1993. The distribution and abundance of Antarctic and subantarctic penguins. SCAR, Scott Polar Research Institute, Cambridge, UK.

Wooller, R. D., Dunlop, J. N., Klomp, N. I., Meathrel, C. E. & Wienecke, B. C. 1991. Seabird abundance, distribution and breeding patterns in relation to the Leeuwin Current. *Jnl Roy Soc WA* 74: 1-4.